General Science

General Science

The Project Gutenberg EBook of General Science, by Bertha M. Clark
This eBook is for the use of anyone anywhere at no cost and with
almost no restrictions whatsoever. You may copy it, give it away or
re-use it under the terms of the Project Gutenberg License included
with this eBook or online at www.gutenberg.net

Title: General Science

Author: Bertha M. Clark

Release Date: August 25, 2005 [EBook #16593]

Language: English

Character set encoding: ISO-8859-1

*** START OF THIS PROJECT GUTENBERG EBOOK GENERAL SCIENCE ***

Produced by John Hagerson, Kevin Handy, Sankar Viswanathan and the Online Distributed Proofreading Team at http://www.pgdp.net

GENERAL SCIENCE

BY

BERTHA M. CLARK, PH.D.

HEAD OF THE SCIENCE DEPARTMENT

WILLIAM PENN HIGH SCHOOL FOR GIRLS, PHILADELPHIA

NEW YORK - CINCINNATI - CHICAGO

AMERICAN BOOK COMPANY

1912

PREFACE

This book is not intended to prepare for college entrance examinations; it will not, in fact, prepare for any of the present-day stock examinations in physics, chemistry, or hygiene, but it should prepare the thoughtful reader to meet wisely and actively some of life's important problems, and should enable him to pass muster on the principles and theories underlying scientific, and therefore economic, management, whether in the shop or in the home.

We hear a great deal about the conservation of our natural resources, such as forests and waterways; it is hoped that this book will show the vital importance of the conservation of human strength and health, and the irreparable loss to society of energy uselessly dissipated, either in idle worry or in aimless activity. Most of us would reproach ourselves for lack of shrewdness if we spent for any article more than it was worth, yet few of us consider that we daily expend on domestic and business tasks an amount of energy far in excess of that actually required. The farmer who flails his grain instead of threshing it wastes time and energy; the housewife who washes with her hands alone and does not aid herself by the use of washing machine and proper bleaching agents dissipates energy sadly needed for other duties.

The Chapter on machines is intended not only as a stimulus to the invention of further labor-saving devices, but also as an eye opener to those who, in the future struggle for existence, must perforce go to the wall unless they understand how to make use of contrivances whereby man's limited physical strength is made effective for larger tasks.

The Chapter on musical instruments is more detailed than seems warranted at first sight; but interest in orchestral instruments is real and general, and there is a persistent desire for intelligent information relative to musical instruments. The child of the laborer as well as the child of the merchant finds it possible to attend some of the weekly orchestral concerts, with their tiers of cheap seats, and nothing adds more to the enjoyment and instruction of such hours than an intimate acquaintance with the leading instruments. Unless this is given in the public schools, a large percentage of mankind is deprived of it, and it is for this reason that so large a share of the treatment of sound has been devoted to musical instruments.

The treatment of electricity is more theoretical than that used in preceding Chapters, but the subject does not lend itself readily to popular presentation; and, moreover, it is assumed that the information and training acquired in the previous work will give the pupil power to understand the more advanced thought and method.

The real value of a book depends not so much upon the information given as upon the permanent interest stimulated and the initiative aroused. The youthful mind, and indeed the average adult mind as well, is singularly non-logical and incapable of continued concentration, and loses interest under too consecutive thought and sustained style. For this reason the author has sacrificed at times detail to general effect, logical development to present-day interest and facts, and has made use of a popular, light style of writing as well as of the more formal and logical style common to books of science.

No claim is made to originality in subject matter. The actual facts, theories, and principles used are such as have been presented in previous textbooks of science, but the manner and sequence of presentation are new and, so far as I know, untried elsewhere. These are such as in my experience have aroused the greatest interest and initiative, and such as have at the same time given the maximum benefit from the informational standpoint. In no case, however, is mental training sacrificed to information; but mental development is sought through the student's willing and interested participation in the actual daily happenings of the home and the shop and the field, rather than through formal recitations and laboratory experiments.

Practical laboratory work in connection with the study of this book is provided for in my *Laboratory Manual in General Science*, which contains directions for a series of experiments designed to make the pupil familiar with the facts and theories discussed in the textbook.

I have sought and have gained help from many of the standard textbooks, new and old. The following firms have kindly placed cuts at my disposal, and have thus materially aided in the preparation of the illustrations: American Radiator Company; Commercial Museum, Philadelphia; General Electric Company; Hershey Chocolate Company; *Scientific American*; The Goulds Manufacturing Company; Victor Talking Machine Company. Acknowledgment is also due to Professor Alvin Davison for figures 19, 23, 29, 142, and 161.

Mr. W.D. Lewis, Principal of the William Penn High School, has read the manuscript and has given me the benefit of his experience and interest. Miss. Helen Hill, librarian of the same school, has been of invaluable service as regards suggestions and proof reading. Miss. Droege, of the Baldwin School, Bryn Mawr, has also been of very great service. Practically all of my assistants have given of their time and skill to the preparation of the work, but the list is too long for individual mention.

BERTHA M. CLARK.

WILLIAM PENN HIGH SCHOOL.

CONTENTS

CHAPTER

I.

HEAT

II. TEMPERATURE AND HEAT

III. OTHER FACTS ABOUT HEAT

IV. BURNING OR OXIDATION

V. FOOD

VI. WATER

VII. AIR

VIII. GENERAL PROPERTIES OF GASES

IX. INVISIBLE OBJECTS

X. LIGHT

XI. REFRACTION

XII. PHOTOGRAPHY

XIII. COLOR

XIV. HEAT AND LIGHT AS COMPANIONS

XV. ARTIFICIAL LIGHTING

XVI. MAN'S WAY OF HELPING HIMSELF

CHAPTER

XVII. THE POWER BEHIND THE ENGINE

XVIII. PUMPS AND THEIR VALUE TO MAN

XIX. THE WATER PROBLEM OF A LARGE CITY

XX. MAN'S CONQUEST OF SUBSTANCES

XXI. FERMENTATION

XXII. BLEACHING

XXIII. DYEING

XXIV. CHEMICALS AS DISINFECTANTS AND PRESERVATIVES

XXV. DRUGS AND PATENT MEDICINES

XXVI. NITROGEN AND ITS RELATION TO PLANTS

XXVII. SOUND

XXVIII. MUSICAL INSTRUMENTS

XXIX. SPEAKING AND HEARING

XXX. ELECTRICITY

XXXI. SOME USES OF ELECTRICITY

XXXII. MODERN ELECTRICAL INVENTIONS

XXXIII. MAGNETS AND CURRENTS

XXXIV. HOW ELECTRICITY MAY BE MEASURED

XXXV. HOW ELECTRICITY IS OBTAINED ON A LARGE SCALE

INDEX

CHAPTER I

GENERAL SCIENCE

CHAPTER I

HEAT

1. **Value of Fire.** Every day, uncontrolled fire wipes out human lives and destroys vast amounts of property; every day, fire, controlled and regulated in stove and furnace, cooks our food and warms our houses. Fire melts ore and allows of the forging of iron, as in the blacksmith's shop, and of the fashioning of innumerable objects serviceable to man. Heated boilers change water into the steam which drives our engines on land and sea. Heat causes rain and wind, fog and cloud; heat enables vegetation to grow and thus indirectly provides our food. Whether heat comes directly from the sun or from artificial sources such as coal, wood, oil, or electricity, it is vitally connected with our daily life, and for this reason the facts and theories relative to it are among the most important that can be studied. Heat, if properly regulated and controlled, would never be injurious to man; hence in the following paragraphs heat will be considered merely in its helpful capacity.

2. **General Effect of Heat.** *Expansion and Contraction.* One of the best-known effects of heat is the change which it causes in the size of a substance. Every housewife knows that if a kettle is filled with cold water to begin with, there will be an overflow as soon as the water becomes heated. Heat causes not only water, but all other liquids, to occupy more space, or to expand, and in some cases the expansion, or increase in size, is surprisingly large. For example, if 100 pints of ice water is heated in a kettle, the 100 pints will steadily expand until, at the boiling point, it will occupy as much space as 104 pints of ice water.

The expansion of water can be easily shown by heating a flask (Fig. I) filled with water and closed by a cork through which a narrow tube passes. As the water is heated, it expands and forces its way up the narrow tube. If the heat is removed, the liquid cools, contracts, and

CHAPTER I

slowly falls in the tube, resuming in time its original size or volume. A similar observation can be made with alcohol, mercury, or any other convenient liquid.

[Illustration: FIG. 1.--As the water becomes warmer it expands and rise in the narrow tube.]

Not only liquids are affected by heat and cold, but solids also are subject to similar changes. A metal ball which when cool will just slip through a ring (Fig. 2) will, when heated, be too large to slip through the ring. Telegraph and telephone wires which in winter are stretched taut from pole to pole, sag in hot weather and are much too long. In summer they are exposed to the fierce rays of the sun, become strongly heated, and expand sufficiently to sag. If the wires were stretched taut in the summer, there would not be sufficient leeway for the contraction which accompanies cold weather, and in winter they would snap.

[Illustration: FIG. 2--When the ball is heated, it become too large to slip through the ring.]

Air expands greatly when heated (Fig. 3), but since air is practically invisible, we are not ordinarily conscious of any change in it. The expansion of air can be readily shown by putting a drop of ink in a thin glass tube, inserting the tube in the cork of a flask, and applying heat to the flask (Fig. 4). The ink is forced up the tube by the expanding air. Even the warmth of the hand is generally sufficient to cause the drop to rise steadily in the tube. The rise of the drop of ink shows that the air in the flask occupies more space than formerly, and since the quantity of air has not changed, each cubic inch of space must hold less warm air than| it held of cold air; that is, one cubic inch of warm air weighs less than one cubic inch of cold air, or warm air is less dense than cold air. All gases, if not confined, expand when heated and contract as they cool. Heat, in general, causes substances to expand or become less dense.

[Illustration: FIG. 3--As the air in *A* is heated, it expands and escapes in the form of bubbles.]

3. **Amount of Expansion and Contraction.** While most substances expand when heated and contract when cooled, they are not all affected equally by the same changes in temperature. Alcohol expands more than water, and water more than mercury. Steel wire which measures 1/4 mile on a snowy day will gain 25 inches in length on a warm summer day, and an aluminum wire under the same conditions would gain 50 inches in length.

[Illustration: FIG. 4.--As the air in *A* is heated, it expands and forces the drop of ink up the tube.]

4. **Advantages and Disadvantages of Expansion and Contraction.** We owe the snug fit of metal tires and bands to the expansion and contraction resulting from heating and cooling. The tire of a wagon wheel is made slightly smaller than the wheel which it is to protect; it is then put into a very hot fire and heated until it has expanded sufficiently to slip on the wheel. As the tire cools it contracts and fits the wheel closely.

In a railroad, spaces are usually left between consecutive rails in order to allow for expansion during the summer.

The unsightly cracks and humps in cement floors are sometimes due to the expansion resulting from heat (Fig. 5). Cracking from this cause can frequently be avoided by cutting the soft cement into squares, the spaces between them giving opportunity for expansion just as do the spaces between the rails of railroads.

[Illustration: FIG. 5: A cement walk broken by expansion due to sun heat.]

In the construction of long wire fences provision must be made for tightening the wire in summer, otherwise great sagging would occur.

Heat plays an important part in the splitting of rocks and in the formation of débris. Rocks in exposed places are greatly affected by changes in temperature, and in regions where the changes in temperature are sudden, severe, and frequent, the rocks are not able to withstand the strain of expansion and contraction, and as a result

CHAPTER I 10

crack and split. In the Sahara Desert much crumbling of the rock into sand has been caused by the intense heat of the day followed by the sharp frost of night. The heat of the day causes the rocks to expand, and the cold of night causes them to contract, and these two forces constantly at work loosen the grains of the rock and force them out of place, thus producing crumbling.

[Illustration: FIG. 6.--Splitting and crumbling of rock caused by alternating heat and cold.]

The surface of the rock is the most exposed part, and during the day the surface, heated by the sun's rays, expands and becomes too large for the interior, and crumbling and splitting result from the strain. With the sudden fall of temperature in the late afternoon and night, the surface of the rock becomes greatly chilled and colder than the rock beneath; the surface rock therefore contracts and shrinks more than the underlying rock, and again crumbling results (Fig. 6).

[Illustration: FIG. 7.--Debris formed from crumbled rock.]

On bare mountains, the heating and cooling effects of the sun are very striking(Fig. 7); the surface of many a mountain peak is covered with cracked rock so insecure that a touch or step will dislodge the fragments and start them down the mountain slope. The lower levels of mountains are frequently buried several feet under débris which has been formed in this way from higher peaks, and which has slowly accumulated at the lower levels.

5. Temperature. When an object feels hot to the touch, we say that it has a high temperature; when it feels cold to the touch, that it has a low temperature; but we are not accurate judges of heat. Ice water seems comparatively warm after eating ice cream, and yet we know that ice water is by no means warm. A room may seem warm to a person who has been walking in the cold air, while it may feel decidedly cold to some one who has come from a warmer room. If the hand is cold, lukewarm water feels hot, but if the hand has been in very hot water and is then transferred to lukewarm water, the latter will seem cold. We see that the sensation or feeling of warmth is not an accurate guide to the temperature of a substance; and yet until 1592,

CHAPTER I

one hundred years after the discovery of America, people relied solely upon their sensations for the measurement of temperature. Very hot substances cannot be touched without injury, and hence inconvenience as well as the necessity for accuracy led to the invention of the thermometer, an instrument whose operation depends upon the fact that most substances expand when heated and contract when cooled.

[Illustration: FIG. 8.--Making a thermometer.]

6. The Thermometer. The modern thermometer consists of a glass tube at the lower end of which is a bulb filled with mercury or colored alcohol (Fig. 8). After the bulb has been filled with the mercury, it is placed in a beaker of water and the water is heated by a Bunsen burner. As the water becomes warmer and warmer the level of the mercury in the tube steadily rises until the water boils, when the level remains stationary (Fig. 9). A scratch is made on the tube to indicate the point to which the mercury rises when the bulb is placed in boiling water, and this point is marked 212°. The tube is then removed from the boiling water, and after cooling for a few minutes, it is placed in a vessel containing finely chopped ice (Fig. 10). The mercury column falls rapidly, but finally remains stationary, and at this level another scratch is made on the tube and the point is marked 32°. The space between these two points, which represent the temperatures of boiling water and of melting ice, is divided into 180 equal parts called degrees. The thermometer in use in the United States is marked in this way and is called the Fahrenheit thermometer after its designer. Before the degrees are etched on the thermometer the open end of the tube is sealed.

[Illustration: FIG. 9.--Determining one of the fixed points of a thermometer.]

The Centigrade thermometer, in use in foreign countries and in all scientific work, is similar to the Fahrenheit except that the fixed points are marked 100° and 0°, and the interval between the points is divided into 100 equal parts instead of into 180.

The boiling point of water is 212° F. or 100° C.

The melting point of ice is 32° F. or 0° C.

Glass thermometers of the above type are the ones most generally used, but there are many different types for special purposes.

[Illustration: FIG. 10.--Determining the lower fixed point of a thermometer.]

7. Some Uses of a Thermometer. One of the chief values of a thermometer is the service it has rendered to medicine. If a thermometer is held for a few minutes under the tongue of a normal, healthy person, the mercury will rise to about 98.4° F. If the temperature of the body registers several degrees above or below this point, a physician should be consulted immediately. The temperature of the body is a trustworthy indicator of general physical condition; hence in all hospitals the temperature of patients is carefully taken at stated intervals.

Commercially, temperature readings are extremely important. In sugar refineries the temperature of the heated liquids is observed most carefully, since a difference in temperature, however slight, affects not only the general appearance of sugars and sirups, but the quality as well. The many varieties of steel likewise show the influence which heat may have on the nature of a substance. By observation and tedious experimentation it has been found that if hardened steel is heated to about 450° F. and quickly cooled, it gives the fine cutting edge of razors; if it is heated to about 500° F. and then cooled, the metal is much coarser and is suitable for shears and farm implements; while if it is heated but 50° F. higher, that is, to 550° F., it gives the fine elastic steel of watch springs.

[Illustration: FIG. 11.--A well-made commercial thermometer.]

A thermometer could be put to good use in every kitchen; the inexperienced housekeeper who cannot judge of the "heat" of the oven would be saved bad bread, etc., if the thermometer were a part of her equipment. The thermometer can also be used in detecting adulterants. Butter should melt at 94° F.; if it does not, you may be sure that it is adulterated with suet or other cheap fat. Olive oil should

CHAPTER I 13

be a clear liquid above 75° F.; if, above this temperature, it looks cloudy, you may be sure that it too is adulterated with fat.

8. Methods of Heating Buildings. *Open Fireplaces and Stoves.* Before the time of stoves and furnaces, man heated his modest dwelling by open fires alone. The burning logs gave warmth to the cabin and served as a primitive cooking agent; and the smoke which usually accompanies burning bodies was carried away by means of the chimney. But in an open fireplace much heat escapes with the smoke and is lost, and only a small portion streams into the room and gives warmth.

When fuel is placed in an open fireplace (Fig. 12) and lighted, the air immediately surrounding the fire becomes warmer and, because of expansion, becomes lighter than the cold air above. The cold air, being heavier, falls and forces the warmer air upward, and along with the warm air goes the disagreeable smoke. The fall of the colder and heavier air, and the rise of the warmer and hence lighter air, is similar to the exchange which takes place when water is poured on oil; the water, being heavier than oil, sinks to the bottom and forces the oil to the surface. The warmer air which escapes up the chimney carries with it the disagreeable smoke, and when all the smoke is got rid of in this way, the chimney is said to draw well.

[Illustration: FIG. 12.--The open fireplace as an early method of heating.]

As the air is heated by the fire it expands, and is pushed up the chimney by the cold air which is constantly entering through loose windows and doors. Open fireplaces are very healthful because the air which is driven out is impure, while the air which rushes in is fresh and brings oxygen to the human being.

But open fireplaces, while pleasant to look at, are not efficient for either heating or cooking. The possibilities for the latter are especially limited, and the invention of stoves was a great advance in efficiency, economy, and comfort. A stove is a receptacle for fire, provided with a definite inlet for air and a definite outlet for smoke, and able to radiate into the room most of the heat produced from the fire which burns within. The inlet, or draft, admits enough air to cause the fire to burn brightly or slowly as the case may be. If we wish a hot fire, the draft is opened wide and enough air enters to produce a strong glow. If we wish a low fire, the inlet is only partially opened, and just enough air enters to keep the fuel smoldering.

When the fire is started, the damper should be opened wide in order to allow the escape of smoke; but after the fire is well started there is less smoke, and the damper may be partly closed. If the damper is kept open, coal is rapidly consumed, and the additional heat passes out through the chimney, and is lost to use.

9. Furnaces. *Hot Air.* The labor involved in the care of numerous stoves is considerable, and hence the advent of a central heating stove, or furnace, was a great saving in strength and fuel. A furnace is a stove arranged as in Figure 13. The stove S, like all other stoves, has an inlet for air and an outlet C for smoke; but in addition, it has built around it a chamber in which air circulates and is warmed. The air warmed by the stove is forced upward by cold air which enters from outside. For example, cold air constantly entering at E drives the air heated by S through pipes and ducts to the rooms to be heated.

The metal pipes which convey the heated air from the furnace to the ducts are sometimes covered with felt, asbestos, or other non-conducting material in order that heat may not be lost during transmission. The ducts which receive the heated air from the pipes are built in the non-conducting walls of the house, and hence lose practically no heat. The air which reaches halls and rooms is therefore warm, in spite of its long journey from the cellar.

[Illustration: FIG. 13.--A furnace. Pipes conduct hot air to the rooms.]

Not only houses are warmed by a central heating stove, but whole communities sometimes depend upon a central heating plant. In the latter case, pipes closely wrapped with a non-conducting material carry steam long distances underground to heat remote buildings. Overbrook and Radnor, Pa., are towns in which such a system is used.

CHAPTER I

10. **Hot-water Heating.** The heated air which rises from furnaces is seldom hot enough to warm large buildings well; hence furnace heating is being largely supplanted by hot-water heating.

The principle of hot-water heating is shown by the following simple experiment. Two flasks and two tubes are arranged as in Figure 15, the upper flask containing a colored liquid and the lower flask clear water. If heat is applied to B, one can see at the end of a few seconds the downward circulation of the colored liquid and the upward circulation of the clear water. If we represent a boiler by B, a radiator by the coiled tube, and a safety tank by C, we shall have a very fair illustration of the principle of a hot-water heating system. The hot water in the radiators cools and, in cooling, gives up its heat to the rooms and thus warms them.

[Illustration: FIG. 14.--Hot-water heating.]

In hot-water heating systems, fresh air is not brought to the rooms, for the radiators are closed pipes containing hot water. It is largely for this reason that thoughtful people are careful to raise windows at intervals. Some systems of hot-water heating secure ventilation by confining the radiators to the basement, to which cold air from outside is constantly admitted in such a way that it circulates over the radiators and becomes strongly heated. This warm fresh air then passes through ordinary flues to the rooms above.

[Illustration: FIG. 15.--The principle of hot-water heating.]

In Figure 16, a radiator is shown in a boxlike structure in the cellar. Fresh air from outside enters a flue at the right, passes the radiator, where it is warmed, and then makes its way to the room through a flue at the left. The warm air which thus enters the room is thoroughly fresh. The actual labor involved in furnace heating and in hot-water heating is practically the same, since coal must be fed to the fire, and ashes must be removed; but the hot-water system has the advantage of economy and cleanliness.

[Illustration: FIG. 16.--Fresh air from outside circulates over the radiators and then rises into the rooms to be heated.]

11. Fresh Air. Fresh air is essential to normal healthy living, and 2000 cubic feet of air per hour is desirable for each individual. If a gentle breeze is blowing, a barely perceptible opening of a window will give the needed amount, even if there are no additional drafts of fresh air into the room through cracks. Most houses are so loosely constructed that fresh air enters imperceptibly in many ways, and whether we will or no, we receive some fresh air. The supply is, however, never sufficient in itself and should not be depended upon alone. At night, or at any other time when gas lights are required, the need for ventilation increases, because every gas light in a room uses up the same amount of air as four people.

[Illustration: FIG. 17.--The air which goes to the schoolrooms is warmed by passage over the radiators.]

In the preceding Section, we learned that many houses heated by hot water are supplied with fresh-air pipes which admit fresh air into separate rooms or into suites of rooms. In some cases the amount which enters is so great that the air in a room is changed three or four times an hour. The constant inflow of cold air and exit of warm air necessitates larger radiators and more hot water and hence more coal to heat the larger quantity of water, but the additional expense is more than compensated by the gain in health.

12. Winds and Currents. The gentlest summer breezes and the fiercest blasts of winter are produced by the unequal heating of air. We have seen that the air nearest to a stove or hot object becomes hotter than the adjacent air, that it tends to expand and is replaced and pushed upward and outward by colder, heavier air falling downward. We have learned also that the moving liquid or gas carries with it heat which it gradually gives out to surrounding bodies.

When a liquid or a gas moves away from a hot object, carrying heat with it, the process is called *convection*.

Convection is responsible for winds and ocean currents, for land and sea breezes, and other daily phenomena.

CHAPTER I

The Gulf Stream illustrates the transference of heat by convection. A large body of water is strongly heated at the equator, and then moves away, carrying heat with it to distant regions, such as England and Norway.

Owing to the shape of the earth and its position with respect to the sun, different portions of the earth are unequally heated. In those portions where the earth is greatly heated, the air likewise will be heated; there will be a tendency for the air to rise, and for the cold air from surrounding regions to rush in to fill its place. In this way winds are produced. There are many circumstances which modify winds and currents, and it is not always easy to explain their direction and velocity, but one very definite cause is the unequal heating of the surface of the earth.

13. Conduction. A poker used in stirring a fire becomes hot and heats the hand grasping the poker, although only the opposite end of the poker has actually been in the fire. Heat from the fire passed into the poker, traveled along it, and warmed it. When heat flows in this way from a warm part of a body to a colder part, the process is called *conduction*. A flatiron is heated by conduction, the heat from the warm stove passing into the cold flatiron and gradually heating it.

In convection, air and water circulate freely, carrying heat with them; in conduction, heat flows from a warm region toward a cold region, but there is no apparent motion of any kind.

Heat travels more readily through some substances than through others. All metals conduct heat well; irons placed on the fire become heated throughout and cannot be grasped with the bare hand; iron utensils are frequently made with wooden handles, because wood is a poor conductor and does not allow heat from the iron to pass through it to the hand. For the same reason a burning match may be held without discomfort until the flame almost reaches the hand.

Stoves and radiators are made of metal, because metals conduct heat readily, and as fast as heat is generated within the stove by the burning of fuel, or introduced into the radiator by the hot water, the heat is conducted through the metal and escapes into the room.

Hot-water pipes and steam pipes are usually wrapped with a non-conducting substance, or insulator, such as asbestos, in order that the heat may not escape, but shall be retained within the pipes until it reaches the radiators within the rooms.

The invention of the "Fireless Cooker" depended in part upon the principle of non-conduction. Two vessels, one inside the other, are separated by sawdust, asbestos, or other poor conducting material (Fig. 18). Foods are heated in the usual way to the boiling point or to a high temperature, and are then placed in the inner vessel. The heat of the food cannot escape through the non-conducting material which surrounds it, and hence remains in the food and slowly cooks it.

[Illustration: FIG. 18.--A fireless cooker.]

A very interesting experiment for the testing of the efficacy of non-conductors may be easily performed. Place hot water in a metal vessel, and note by means of a thermometer the *rapidity* with which the water cools; then place water of the same temperature in a second metal vessel similar to the first, but surrounded by asbestos or other non-conducting material, and note the *slowness* with which the temperature falls.

Chemical Change, an Effect of Heat. This effect of heat has a vital influence on our lives, because the changes which take place when food is cooked are due to it. The doughy mass which goes into the oven, comes out a light spongy loaf; the small indigestible rice grain comes out the swollen, fluffy, digestible grain. Were it not for the chemical changes brought about by heat, many of our present foods would be useless to man. Hundreds of common materials like glass, rubber, iron, aluminum, etc., are manufactured by processes which involve chemical action caused by heat.

CHAPTER II

TEMPERATURE AND HEAT

14. **Temperature not a Measure of the Amount of Heat Present.** If two similar basins containing unequal quantities of water are placed in the sunshine on a summer day, the smaller quantity of water will become quite warm in a short period of time, while the larger quantity will become only lukewarm. Both vessels receive the same amount of heat from the sun, but in one case the heat is utilized in heating to a high temperature a small quantity of water, while in the second case the heat is utilized in warming to a lower degree a larger quantity of water. Equal amounts of heat do not necessarily produce equivalent temperatures, and equal temperatures do not necessarily indicate equal amounts of heat. It takes more heat to raise a gallon of water to the boiling point than it does to raise a pint of water to the boiling point, but a thermometer would register the same temperature in the two cases. The temperature of boiling water is 100° C. whether there is a pint of it or a gallon. Temperature is independent of the quantity of matter present; but the amount of heat contained in a substance at any temperature is not independent of quantity, being greater in the larger quantity.

15. **The Unit of Heat.** It is necessary to have a unit of heat just as we have a unit of length, or a unit of mass, or a unit of time. One unit of heat is called a *calorie*, and is the amount of heat which will change the temperature of 1 gram of water 1° C. It is the amount of heat given out by 1 gram of water when its temperature falls 1° C., or the amount of heat absorbed by 1 gram of water when its temperature rises 1° C. If 400 grams of water are heated from 0° to 5° C., the amount of heat which has entered the water is equivalent to 5 × 400 or 2000 calories; if 200 grams of water cool from 25° to 20° C., the heat given out by the water is equivalent to 5 × 200 or 1000 calories.

16. **Some Substances Heat more readily than Others.** If two equal quantities of water at the same temperature are exposed to the sun for the same length of time, their final temperatures will be the same. If, however, equal quantities of different substances are exposed, the temperatures resulting from the heating will not necessarily be the same. If a basin containing 1 lb. of mercury is put on the fire, side by side with a basin containing an equal quantity of water, the temperatures of the two substances will vary greatly at the end of a short time. The mercury will have a far higher temperature than the

water, in spite of the fact that the amount of mercury is as great as the amount of water and that the heat received from the fire has been the same in each case. Mercury is not so difficult to heat as water; less heat being required to raise its temperature 1° than is required to raise the temperature of an equal quantity of water 1°. In fact, mercury is 30 times as easy to heat as water, and it requires only one thirtieth as much fire to heat a given quantity of mercury 1° as to heat the same quantity of water 1°.

17. Specific Heat. We know that different substances are differently affected by heat. Some substances, like water, change their temperature slowly when heated; others, like mercury, change their temperature very rapidly when heated. The number of calories needed by 1 gram of a substance in order that its temperature may be increased 1° C. is called the *specific heat* of a substance; or, specific heat is the number of calories given out by 1 gram of a substance when its temperature falls 1° C. For experiments on the determination of specific heat, see Laboratory Manual.

Water has the highest specific heat of any known substance except hydrogen; that is, it requires more heat to raise the temperature of water a definite number of degrees than it does to raise the temperature of an equal amount of any other substance the same number of degrees. Practically this same thing can be stated in another way: Water in cooling gives out more heat than any other substance in cooling through the same number of degrees. For this reason water is used in foot warmers and in hot-water bags. If a copper lid were used as a foot warmer, it would give the feet only .095 as much heat as an equal weight of water; a lead weight only .031 as much heat as water. Flatirons are made of iron because of the relatively high specific heat of iron. The flatiron heats slowly and cools slowly, and, because of its high specific heat, not only supplies the laundress with considerable heat, but eliminates for her the frequent changing of the flatiron.

18. Water and Weather. About four times as much heat is required to heat a given quantity of water one degree as to heat an equal quantity of earth. In summer, when the rocks and the sand along the shore are burning hot, the ocean and lakes are pleasantly cool, although the

amount of heat present in the water is as great as that present in the earth. In winter, long after the rocks and sand have given out their heat and have become cold, the water continues to give out the vast store of heat accumulated during the summer. This explains why lands situated on or near large bodies of water usually have less variation in temperature than inland regions. In the summer the water cools the region; in the winter, on the contrary, the water heats the region, and hence extremes of temperature are practically unknown.

19. Sources of Heat. Most of the heat which we enjoy and use we owe to the sun. The wood which blazes on the hearth, the coal which glows in the furnace, and the oil which burns in the stove owe their existence to the sun.

Without the warmth of the sun seeds could not sprout and develop into the mighty trees which yield firewood. Even coal, which lies buried thousands of feet below the earth's surface, owes its existence in part to the sun. Coal is simply buried vegetation,--vegetation which sprouted and grew under the influence of the sun's warm rays. Ages ago trees and bushes grew "thick and fast," and the ground was always covered with a deep layer of decaying vegetable matter. In time some of this vast supply sank into the moist soil and became covered with mud. Then rock formed, and the rock pressed down upon the sunken vegetation. The constant pressure, the moisture in the ground, and heat affected the underground vegetable mass, and slowly changed it into coal.

The buried forest and thickets were not all changed into coal. Some were changed into oil and gas. Decaying animal matter was often mixed with the vegetable mass. When the mingled animal and vegetable matter sank into moist earth and came under the influence of pressure, it was slowly changed into oil and gas.

The heat of our bodies comes from the foods which we eat. Fruits, grain, etc., could not grow without the warmth and the light of the sun. The animals which supply our meats likewise depend upon the sun for light and warmth.

The sun, therefore, is the great source of heat; whether it is the heat which comes directly from the sun and warms the atmosphere, or the heat which comes from burning coal, wood, and oil.

CHAPTER III

OTHER FACTS ABOUT HEAT

20. Boiling. *Heat absorbed in Boiling.* If a kettle of water is placed above a flame, the temperature of the water gradually increases, and soon small bubbles form at the bottom of the kettle and begin to rise through the water. At first the bubbles do not get far in their ascent, but disappear before they reach the surface; later, as the water gets hotter and hotter, the bubbles become larger and more numerous, rise higher and higher, and finally reach the surface and pass from the water into the air; steam comes from the vessel, and the water is said to *boil.* The temperature at which a liquid boils is called the boiling point.

While the water is heating, the temperature steadily rises, but as soon as the water begins to boil the thermometer reading becomes stationary and does not change, no matter how hard the water boils and in spite of the fact that heat from the flame is constantly passing into the water.

If the flame is removed from the boiling water for but a second, the boiling ceases; if the flame is replaced, the boiling begins again immediately. Unless heat is constantly supplied, water at the boiling point cannot be transformed into steam.

The number of calories which must be supplied to 1 gram of water at the boiling point in order to change it into steam at the same temperature is called the heat of vaporization; it is the heat necessary to change 1 gram of water at the boiling point into steam of the same temperature.

21. The Amount of Heat Absorbed. The amount of heat which must be constantly supplied to water at the boiling point in order to change it

into steam is far greater than we realize. If we put a beaker of ice water (water at 0° C.) over a steady flame, and note (1) the time which elapses before the water begins to boil, and (2) the time which elapses before the boiling water completely boils away, we shall see that it takes about 5-1/4 times as long to change water into steam as it does to change its temperature from 0° C. to 100° C. Since, with a steady flame, it takes 5-1/4 times as long to change water into steam as it does to change its temperature from 0° C. to the boiling point, we conclude that it takes 5-1/4 times as much heat to convert water at the boiling point into steam as it does to raise it from the temperature of ice water to that of boiling water.

The amount of heat necessary to raise the temperature of 1 gram of water 1° C. is equal to 1 calorie, and the amount necessary to raise the temperature 100° C. is equal to 100 calories; hence the amount of heat necessary to convert 1 gram of water at the boiling point into steam at that same temperature is equal to approximately 525 calories. Very careful experiments show the exact heat of vaporization to be 536.1 calories. (See Laboratory Manual.)

22. General Truths. Statements similar to the above hold for other liquids and for solutions. If milk is placed upon a stove, the temperature rises steadily until the boiling point is reached; further heating produces, not a change in temperature, but a change of the water of the milk into steam. As soon as the milk, or any other liquid food, comes to a boil, the gas flame should be lowered until only an occasional bubble forms, because so long as any bubbles form the temperature is that of the boiling point, and further heat merely results in waste of fuel.

We find by experiment that every liquid has its own specific boiling point; for example, alcohol boils at 78° C. and brine at 103° C. Both specific heat and the heat of vaporization vary with the liquid used.

23. Condensation. If one holds a cold lid in the steam of boiling water, drops of water gather on the lid; the steam is cooled by contact with the cold lid and *condenses* into water. Bottles of water brought from a cold cellar into a warm room become covered with a mist of fine drops of water, because the moisture in the air, chilled by contact with the

cold bottles, immediately condenses into drops of water. Glasses filled with ice water show a similar mist.

In Section 21, we saw that 536 calories are required to change 1 gram of water into steam; if, now, the steam in turn condenses into water, it is natural to expect a release of the heat used in transforming water into steam. Experiment shows not only that vapor gives out heat during condensation, but that the amount of heat thus set free is exactly equal to the amount absorbed during vaporization. (See Laboratory Manual.)

We learn that the heat of vaporization is the same whether it is considered as the heat absorbed by 1 gram of water in its change to steam, or as the heat given out by 1 gram of steam during its condensation into water.

24. Practical Application. We understand now the value of steam as a heating agent. Water is heated in a boiler in the cellar, and the steam passes through pipes which run to the various rooms; there the steam condenses into water in the radiators, each gram of steam setting free 536 calories of heat. When we consider the size of the radiators and the large number of grams of steam which they contain, and consider further that each gram in condensing sets free 536 calories, we understand the ease with which buildings are heated by steam.

Most of us have at times profited by the heat of condensation. In cold weather, when there is a roaring fire in the range, the water frequently becomes so hot that it "steams" out of open faucets. If, at such times, the hot water is turned on in a small cold bathroom, and is allowed to run until the tub is well filled, vapor condenses on windows, mirrors, and walls, and the cold room becomes perceptibly warmer. The heat given out by the condensing steam passes into the surrounding air and warms the room.

There is, however, another reason for the rise in temperature. If a large pail of hot soup is placed in a larger pail of cold water, the soup will gradually cool and the cold water will gradually become warmer. A red-hot iron placed on a stand gradually cools, but warms the stand. A hot body loses heat so long as a cooler body is near it; the cold object is heated at the expense of the warmer object, and one loses heat and

CHAPTER III 25

the other gains heat until the temperature of both is the same. Now the hot water in the tub gradually loses heat and the cold air of the room gradually gains heat by convection, but the amount given the room by convection is relatively small compared with the large amount set free by the condensing steam.

25. Distillation. If impure, muddy water is boiled, drops of water will collect on a cold plate held in the path of the steam, but the drops will be clear and pure. When impure water is boiled, the steam from it does not contain any of the impurities because these are left behind in the vessel. If all the water were allowed to boil away, a layer of mud or of other impurities would be found at the bottom of the vessel. Because of this fact, it is possible to purify water in a very simple way. Place over a fire a large kettle closed except for a spout which is long enough to reach across the stove and dip into a bottle. As the liquid boils, steam escapes through the spout, and on reaching the cold bottle condenses and drops into the bottle as pure water. The impurities remain behind in the kettle. Water freed from impurities in this way is called *distilled water*, and the process is called *distillation* (Fig. 19). By this method, the salt water of the ocean may be separated into pure drinking water and salt, and many of the large ocean liners distill from the briny deep all the drinking water used on their ocean voyages.

[Illustration: FIG. 19.--In order that the steam which passes through the coiled tube may be quickly cooled and condensed, cold water is made to circulate around the coil. The condensed steam escapes at *w*.]

Commercially, distillation is a very important process. Turpentine, for example, is made by distilling the sap of pine trees. Incisions are cut in the bark of the long-leaf pine trees, and these serve as channels for the escape of crude resin. This crude liquid is collected in barrels and taken to a distillery, where it is distilled into turpentine and rosin. The turpentine is the product which passes off as vapor, and the rosin is the mass left in the boiler after the distillation of the turpentine.

26. Evaporation. If a stopper is left off a cologne bottle, the contents of the bottle will slowly evaporate; if a dish of water is placed out of doors on a hot day, evaporation occurs very rapidly. The liquids which have

disappeared from the bottle and the dish have passed into the surrounding air in the form of vapor. In Section 20, we saw that water could not pass into vapor without the addition of heat; now the heat necessary for the evaporation of the cologne and water was taken from the air, leaving it slightly cooler. If wet hands are not dried with a towel, but are left to dry by evaporation, heat is taken from the hand in the process, leaving a sensation of coolness. Damp clothing should never be worn, because the moisture in it tends to evaporate at the expense of the bodily heat, and this undue loss of heat from the body produces chills. After a bath the body should be well rubbed, otherwise evaporation occurs at the expense of heat which the body cannot ordinarily afford to lose.

Evaporation is a slow process occurring at all times; it is hastened during the summer, because of the large amount of heat present in the atmosphere. Many large cities make use of the cooling effect of evaporation to lower the temperature of the air in summer; streets are sprinkled not only to lay the dust, but in order that the surrounding air may be cooled by the evaporation of the water.

Some thrifty housewives economize by utilizing the cooling effects of evaporation. Butter, cheese, and other foods sensitive to heat are placed in porous vessels wrapped in wet cloths. Rapid evaporation of the water from the wet cloths keeps the contents of the jars cool, and that without expense other than the muscular energy needed for wetting the cloths frequently.

27. Rain, Snow, Frost, Dew. The heat of the sun causes constant evaporation of the waters of oceans, rivers, streams, and marshes, and the water vapor set free by evaporation passes into the air, which becomes charged with vapor or is said to be humid. Constant, unceasing evaporation of our lakes, streams, and pools would mean a steady decrease in the supply of water available for daily use, if the escaped water were all retained by the atmosphere and lost to the earth. But although the escaped vapor mingles with the atmosphere, hovering near the earth's surface, or rising far above the level of the mountains, it does not remain there permanently. When this vapor meets a cold wind or is chilled in any way, condensation takes place, and a mass of tiny drops of water or of small particles of snow is

formed. When these drops or particles become large enough, they fall to the earth as rain or snow, and in this way the earth is compensated for the great loss of moisture due to evaporation. Fog is formed when vapor condenses near the surface of the earth, and when the drops are so small that they do not fall but hover in the air, the fog is said "not to lift" or "not to clear."

If ice water is poured into a glass, a mist will form on the outside of the glass. This is because the water vapor in the air becomes chilled by contact with the glass and condenses. Often leaves and grass and sidewalks are so cold that the water vapor in the atmosphere condenses on them, and we say a heavy dew has formed. If the temperature of the air falls to the freezing point while the dew is forming, the vapor is frozen and frost is seen instead of dew.

The daily evaporation of moisture into the atmosphere keeps the atmosphere more or less full of water vapor; but the atmosphere can hold only a definite amount of vapor at a given temperature, and as soon as it contains the maximum amount for that temperature, further evaporation ceases. If clothes are hung out on a damp, murky day they do not dry, because the air contains all the moisture it can hold, and the moisture in the clothes has no chance to evaporate. When the air contains all the moisture it can hold, it is said to be saturated, and if a slight fall in temperature occurs when the air is saturated, condensation immediately begins in the form of rain, snow, or fog. If, however, the air is not saturated, a fall in temperature may occur without producing precipitation. The temperature at which air is saturated and condensation begins is called the *dew point*.

28. How Chills are Caused. The discomfort we feel in an overcrowded room is partly due to an excess of moisture in the air, resulting from the breathing and perspiration of many persons. The air soon becomes saturated with vapor and cannot take away the perspiration from our bodies, and our clothing becomes moist and our skin tender. When we leave the crowded "tea" or lecture and pass into the colder, drier, outside air, clothes and skin give up their load of moisture through sudden evaporation. But evaporation requires heat, and this heat is taken from our bodies, and a chill results.

Proper ventilation would eliminate much of the physical danger of social events; fresh, dry air should be constantly admitted to crowded rooms in order to replace the air saturated by the breath and perspiration of the occupants.

29. **Weather Forecasts.** When the air is near the saturation point, the weather is oppressive and is said to be very humid. For comfort and health, the air should be about two thirds saturated. The presence of some water vapor in the air is absolutely necessary to animal and plant life. In desert regions where vapor is scarce the air is so dry that throat trouble accompanied by disagreeable tickling is prevalent; fallen leaves become so dry that they crumble to dust; plants lose their freshness and beauty.

The likelihood of rain or frost is often determined by temperature and humidity. If the air is near saturation and the temperature is falling, it is safe to predict bad weather, because the fall of temperature will probably cause rapid condensation, and hence rain. If, however, the air is not near the saturation point, a fall in temperature will not necessarily produce bad weather.

The measurement of humidity is of far wider importance than the mere forecasting of local weather conditions. The close relation between humidity and health has led many institutions, such as hospitals, schools, and factories, to regulate the humidity of the atmosphere as carefully as they do the temperature. Too great humidity is enervating, and not conducive to either mental or physical exertion; on the other hand, too dry air is equally harmful. In summer the humidity conditions cannot be well regulated, but in winter, when houses are artificially heated, the humidity of a room can be increased by placing pans of water near the registers or on radiators.

30. **Heat Needed to Melt Substances.** If a spoon is placed in a vessel of hot water for a few seconds and then removed, it will be warmer than before it was placed in the hot water. If a lump of melting ice is placed in the vessel of hot water and then removed, the ice will not be warmer than before, but there will be less of it. The heat of the water has been used in melting the ice, not in changing its temperature.

CHAPTER III

If, on a bitter cold day, a pail of snow is brought into a warm room and a thermometer is placed in the snow, the temperature rises gradually until 32° F. is reached, when it becomes stationary, and the snow begins to melt. If the pail is put on the fire, the temperature still remains 32°F., but the snow melts more rapidly. As soon as all the snow is completely melted, however, the temperature begins to rise and rises steadily until the water boils, when it again becomes stationary and remains so during the passage of water into vapor.

We see that heat must be supplied to ice at 0° C. or 32° F. in order to change it into water, and further, that the temperature of the mixture does not rise so long as any ice is present, no matter how much heat is supplied. The amount of heat necessary to melt 1 gram of ice is easily calculated. (See Laboratory Manual.)

Heat must be supplied to ice to melt it. On the other hand, water, in freezing, loses heat, and the amount of heat lost by freezing water is exactly equal to the amount of heat absorbed by melting ice.

The number of units of heat required to melt a unit mass of ice is called the *heat of fusion* of water.

31. Climate. Water, in freezing, loses heat, even though its temperature remains at 0° C. Because water loses heat when it freezes, the presence of large streams of water greatly influences the climate of a region. In winter the heat from the freezing water keeps the temperature of the surrounding higher than it would naturally be, and consequently the cold weather is less severe. In summer water evaporates, heat is taken from the air, and consequently the warm weather is less intense.

32. Molding of Glass and Forging of Iron. The fire which is hot enough to melt a lump of ice may not be hot enough to melt an iron poker; on the other hand, it may be sufficiently hot to melt a tin spoon. Different substances melt, or liquefy, at different temperatures; for example, ice melts at 0° C., and tin at 233° C., while iron requires the relatively high temperature of 1200° C. Most substances have a definite melting or freezing point which never changes so long as the surrounding conditions remain the same.

But while most substances have a definite melting point, some substances do not. If a glass rod is held in a Bunsen burner, it will gradually grow softer and softer, and finally a drop of molten glass will fall from the end of the rod into the fire. The glass did not suddenly become a liquid at a definite temperature; instead it softened gradually, and then melted. While glass is in the soft, yielding, pliable state, it is molded into dishes, bottles, and other useful objects, such as lamp shades, globes, etc. (Fig. 20). If glass melted at a definite temperature, it could not be molded in this way. Iron acts in a similar manner, and because of this property the blacksmith can shape his horseshoes, and the workman can make his engines and other articles of daily service to man.

[Illustration: FIG. 20.--Molten glass being rolled into a form suitable for window panes.]

33. Strange Behavior of Water. One has but to remember that bottles of water burst when they freeze, and that ice floats on water like wood, to know that water expands on freezing or on solidifying. A quantity of water which occupies 100 cubic feet of space will, on becoming ice, need 109 cubic feet of space. On a cold winter night the water sometimes freezes in the water pipes, and the pipes burst. Water is very peculiar in expanding on solidification, because most substances contract on solidifying; gelatin and jelly, for example, contract so much that they shrink from the sides of the dish which contains them.

If water contracted in freezing, ice would be heavier than water and would sink in ponds and lakes as fast as it formed, and our streams and ponds would become masses of solid ice, killing all animal and plant life. But the ice is lighter than water and floats on top, and animals in the water beneath are as free to live and swim as they were in the warm sunny days of summer. The most severe winter cannot freeze a deep lake solid, and in the coldest weather a hole made in the ice will show water beneath the surface. Our ice boats cut and break the ice of the river, and through the water beneath our boats daily ply their way to and fro, independent of winter and its blighting blasts.

While most of us are familiar with the bursting of water pipes on a cold night, few of us realize the influence which freezing water exerts on the character of the land around us.

Water sinks into the ground and, on the approach of winter, freezes, expanding about one tenth of its volume; the expanding ice pushes the earth aside, the force in some cases being sufficient to dislodge even huge rocks. In the early days in New England it was said by the farmers that "rocks grew," because fields cleared of stones in the fall became rock covered with the approach of spring; the rocks and stones hidden underground and unseen in the fall were forced to the surface by the winter's expansion. We have all seen fence posts and bricks pushed out of place because of the heaving of the soil beneath them. Often householders must relay their pavements and walks because of the damage done by freezing water.

The most conspicuous effect of the expansive power of freezing water is seen in rocky or mountainous regions (Fig. 21). Water easily finds entrance into the cracks and crevices of the rocks, where it lodges until frozen; then it expands and acts like a wedge, widening cracks, chiseling off edges, and even breaking rocks asunder. In regions where frequent frosts occur, the destructive action of water works constant changes in the appearance of the land; small cracks and crevices are enlarged, massive rocks are pried up out of position, huge slabs are split off, and particles large and small are forced from the parent rock. The greater part of the debris and rubbish brought down from the mountain slopes by the spring rains owes its origin to the fact that water expands when it freezes.

[Illustration: FIG. 21.--The destruction caused by freezing water.]

34. Heat Necessary to Dissolve a Substance. It requires heat to dissolve any substance, just as it requires heat to change ice to water. If a handful of common salt is placed in a small cup of water and stirred with a thermometer, the temperature of the mixture falls several degrees. This is just what one would expect, because the heat needed to liquefy the salt must come from somewhere, and naturally it comes from the water, thereby lowering the temperature of the water. We know very well that potatoes cease boiling if a pinch of salt is put in the water; this is because the temperature of the water has been lowered by the amount of heat necessary to dissolve the salt.

Let some snow or chopped ice be placed in a vessel and mixed with one third its weight of coarse salt; if then a small tube of cold water is placed in this mixture, the water in the test tube will soon freeze solid. As soon as the snow and salt are mixed they melt. The heat necessary for this comes in part from the air and in part from the water in the test tube, and the water in the tube becomes in consequence cold enough to freeze. But the salt mixture does not freeze because its freezing point is far below that of pure water. The use of salt and ice in ice-cream freezers is a practical application of this principle. The heat necessary for melting the mixture of salt and ice is taken from the cream which thus becomes cold enough to freeze.

CHAPTER IV

BURNING OR OXIDATION

35. Why Things Burn. The heat of our bodies comes from the food we eat; the heat for cooking and for warming our houses comes from coal. The production of heat through the burning of coal, or oil, or gas, or wood, is called combustion. Combustion cannot occur without the presence of a substance called oxygen, which exists rather abundantly in the air; that is, one fifth of our atmosphere consists of this substance which we call oxygen. We throw open our windows to allow fresh air to enter, and we take walks in order to breathe the pure air into our lungs. What we need for the energy and warmth of our bodies is the oxygen in the air. Whether we burn gas or wood or coal, the heat which is produced comes from the power which these various substances possess to combine with oxygen. We open the draft of a stove that it may "draw well": that it may secure oxygen for burning. We throw a blanket over burning material to smother the fire: to keep oxygen away from it. Burning, or oxidation, is combining with oxygen, and the more oxygen you add to a fire, the hotter the fire will burn, and the faster. The effect of oxygen on combustion may be clearly seen by thrusting a smoldering splinter into a jar containing oxygen; the smoldering splinter will instantly flare and blaze, while if it is removed from the jar, it loses its flame and again burns quietly. Oxygen for this experiment can be produced in the following way.

CHAPTER IV

[Illustration: FIG. 22.--Preparing oxygen from potassium chlorate and manganese dioxide.]

36. How to Prepare Oxygen. Mix a small quantity of potassium chlorate with an equal amount of manganese dioxide and place the mixture in a strong test tube. Close the mouth of the tube with a one-hole rubber stopper in which is fitted a long, narrow tube, and clamp the test tube to an iron support, as shown in Figure 22. Fill the trough with water until the shelf is just covered and allow the end of the delivery tube to rest just beneath the hole in the shelf. Fill a medium-sized bottle with water, cover it with a glass plate, invert the bottle in the trough, and then remove the glass plate. Heat the test tube very gently, and when gas bubbles out of the tube, slip the bottle over the opening in the shelf, so that the tube runs into the bottle. The gas will force out the water and will finally fill the bottle. When all the water has been forced out, slip the glass plate under the mouth of the bottle and remove the bottle from the trough. The gas in the bottle is oxygen.

Everywhere in a large city or in a small village, smoke is seen, indicating the presence of fire; hence there must exist a large supply of oxygen to keep all the fires alive. The supply of oxygen needed for the fires of the world comes largely from the atmosphere.

37. Matches. The burning material is ordinarily set on fire by matches, thin strips of wood tipped with sulphur or phosphorus, or both. Phosphorus can unite with oxygen at a fairly low temperature, and if phosphorus is rubbed against a rough surface, the friction produced will raise the temperature of the phosphorus to a point where it can combine with oxygen. The burning phosphorus kindles the wood of the match, and from the burning match the fire is kindled. If you want to convince yourself that friction produces heat, rub a cent vigorously against your coat and note that the cent becomes warm. Matches have been in use less than a hundred years. Primitive man kindled his camp fire by rubbing pieces of dry wood together until they took fire, and this method is said to be used among some isolated distant tribes at the present time. A later and easier way was to strike flint and steel together and to catch the spark thus produced on tinder or dry fungus. Within the memory of some persons now living, the tinder box was a

CHAPTER IV   34

valuable asset to the home, particularly in the pioneer regions of the West.

38. Safety Matches. Ordinary phosphorus, while excellent as a fire-producing material, is dangerously poisonous, and those to whom the dipping of wooden strips into phosphorus is a daily occupation suffer with a terrible disease which usually attacks the teeth and bones of the jaw. The teeth rot and fall out, abscesses form, and bones and flesh begin to decay; the only way to prevent the spread of the disease is to remove the affected bone, and in some instances it has been necessary to remove the entire jaw. Then, too, matches made of yellow or white phosphorus ignite easily, and, when rubbed against any rough surface, are apt to take fire. Many destructive fires have been started by the accidental friction of such matches against rough surfaces.

For these reasons the introduction of the so-called safety match was an important event. When common phosphorus, in the dangerous and easily ignited form, is heated in a closed vessel to about 250° C., it gradually changes to a harmless red mass. The red phosphorus is not only harmless, but it is difficult to ignite, and, in order to be ignited by friction, must be rubbed on a surface rich in oxygen. The head of a safety match is coated with a mixture of glue and oxygen-containing compounds; the surface on which the match is to be rubbed is coated with a mixture of red phosphorus and glue, to which finely powdered glass is sometimes added in order to increase the friction. Unless the head of the match is rubbed on the prepared phosphorus coating, ignition does not occur, and accidental fires are avoided.

Various kinds of safety matches have been manufactured in the last few years, but they are somewhat more expensive than the ordinary form, and hence manufacturers are reluctant to substitute them for the cheaper matches. Some foreign countries, such as Switzerland, prohibit the sale of the dangerous type, and it is hoped that the United States will soon follow the lead of these countries in demanding the sale of safety matches only.

39. Some Unfamiliar Forms of Burning. While most of us think of burning as a process in which flames and smoke occur, there are in

reality many modes of burning accompanied by neither flame nor smoke. Iron, for example, burns when it rusts, because it slowly combines with the oxygen of the air and is transformed into new substances. When the air is dry, iron does not unite with oxygen, but when moisture is present in the air, the iron unites with the oxygen and turns into iron rust. The burning is slow and unaccompanied by the fire and smoke so familiar to us, but the process is none the less burning, or combination with oxygen. Burning which is not accompanied by any of the appearances of ordinary burning is known as oxidation.

The tendency of iron to rust lessens its efficiency and value, and many devices have been introduced to prevent rusting. A coating of paint or varnish is sometimes applied to iron in order to prevent contact with air. The galvanizing of iron is another attempt to secure the same result; in this process iron is dipped into molten zinc, thereby acquiring a coating of zinc, and forming what is known as galvanized iron. Zinc does not combine with oxygen under ordinary circumstances, and hence galvanized iron is immune from rust.

Decay is a process of oxidation; the tree which rots slowly away is undergoing oxidation, and the result of the slow burning is the decomposed matter which we see and the invisible gases which pass into the atmosphere. The log which blazes on our hearth gives out sufficient heat to warm us; the log which decays in the forest gives out an equivalent amount of heat, but the heat is evolved so slowly that we are not conscious of it. Burning accompanied by a blaze and intense heat is a rapid process; burning unaccompanied by fire and appreciable heat is a slow, gradual process, requiring days, weeks, and even long years for its completion.

Another form of oxidation occurs daily in the human body. In Section 35 we saw that the human body is an engine whose fuel is food; the burning of that food in the body furnishes the heat necessary for bodily warmth and the energy required for thought and action. Oxygen is essential to burning, and the food fires within the body are kept alive by the oxygen taken into the body at every breath by the lungs. We see now one reason for an abundance of fresh air in daily life.

40. How to Breathe. Air, which is essential to life and health, should enter the body through the nose and *not through the mouth*. The peculiar nature and arrangement of the membranes of the nose enable the nostrils to clean, and warm, and moisten the air which passes through them to the lungs. Floating around in the atmosphere are dust particles which ought not to get into the lungs. The nose is provided with small hairs and a moist inner membrane which serve as filters in removing solid particles from the air, and in thus purifying it before its entrance into the lungs.

In the immediate neighborhood of three Philadelphia high schools, having an approximate enrollment of over 8000 pupils, is a huge manufacturing plant which day and night pours forth grimy smoke and soot into the atmosphere which must supply oxygen to this vast group of young lives. If the vital importance of nose breathing is impressed upon these young people, the harmful effect of the foul air may be greatly lessened, the smoke particles and germs being held back by the nose filters and never reaching the lungs. If, however, this principle of hygiene is not brought to their attention, the dangerous habit of breathing through the open, or at least partially open, mouth will continue, and objectionable matter will pass through the mouth and find a lodging place in the lungs.

There is another very important reason why nose breathing is preferable to mouth breathing. The temperature of the human body is approximately 98° F., and the air which enters the lungs should not be far below this temperature. If air reaches the lungs through the nose, its journey is relatively long and slow, and there is opportunity for it to be warmed before it reaches the lungs. If, on the other hand, air passes to the lungs by way of the mouth, the warming process is brief and insufficient, and the lungs suffer in consequence. Naturally, the gravest danger is in winter.

41. Cause of Mouth Breathing. Some people find it difficult to breathe through the nostrils on account of growths, called adenoids, in the nose. If you have a tendency toward mouth breathing, let a physician examine your nose and throat.

CHAPTER IV 37

Adenoids not only obstruct breathing and weaken the whole system through lack of adequate air, but they also press upon the blood vessels and nerves of the head and interfere with normal brain development. Moreover, they interfere in many cases with the hearing, and in general hinder activity and growth. The removal of adenoids is simple, and carries with it only temporary pain and no danger. Some physicians claim that the growths disappear in later years, but even if that is true, the physical and mental development of earlier years is lost, and the person is backward in the struggle for life and achievement.

[Illustration: FIG. 23.--Intelligent expression is often lacking in children with adenoid growths.]

42. How to Build a Fire. Substances differ greatly as to the ease with which they may be made to burn or, in technical terms, with which they may be made to unite with oxygen. For this reason, we put light materials, like shavings, chips, and paper, on the grate, twisting the latter and arranging it so that air (oxygen in the air) can reach a large surface; upon this we place small sticks of wood, piling them across each other so as to allow entrance for the oxygen; and finally upon this we place our hard wood or coal.

The coal and the large sticks cannot be kindled with a match, but the paper and shavings can, and these in burning will heat the large sticks until they take fire and in turn kindle the coal.

43. Spontaneous Combustion. We often hear of fires "starting themselves," and sometimes the statement is true. If a pile of oily rags is allowed to stand for a time, the oily matter will begin to combine slowly with oxygen and as a result will give off heat. The heat thus given off is at first insufficient to kindle a fire; but as the heat is retained and accumulated, the temperature rises, and finally the kindling point is reached and the whole mass bursts into flames. For safety's sake, all oily cloths should be burned or kept in metal vessels.

44. The Treatment of Burns. In spite of great caution, burns from fires, steam, or hot water do sometimes occur, and it is well to know how to relieve the suffering caused by them and how to treat the injury in

order to insure rapid healing.

Burns are dangerous because they destroy skin and thus open up an entrance into the body for disease germs, and in addition because they lay bare nerve tissue which thereby becomes irritated and causes a shock to the entire system.

In mild burns, where the skin is not broken but is merely reddened, an application of moist baking soda brings immediate relief. If this substance is not available, flour paste, lard, sweet oil, or vaseline may be used.

In more severe burns, where blisters are formed, the blisters should be punctured with a sharp, sterilized needle and allowed to discharge their watery contents before the above remedies are applied.

In burns severe enough to destroy the skin, disinfection of the open wound with weak carbolic acid or hydrogen peroxide is very necessary. After this has been done, a soft cloth soaked in a solution of linseed oil and limewater should be applied and the whole bandaged. In such a case, it is important not to use cotton batting, since this sticks to the rough surface and causes pain when removed.

45. Carbon Dioxide. *A Product of Burning.* When any fuel, such as coal, gas, oil, or wood, burns, it sends forth gases into the surrounding atmosphere. These gases, like air, are invisible, and were unknown to us for a long time. The chief gas formed by a burning substance is called carbon dioxide (CO_2) because it is composed of one part of carbon and two parts of oxygen. This gas has the distinction of being the most widely distributed gaseous compound of the entire world; it is found in the ocean depths and on the mountain heights, in brilliantly lighted rooms, and most abundantly in manufacturing towns where factory chimneys constantly pour forth hot gases and smoke.

Wood and coal, and in fact all animal and vegetable matter, contain carbon, and when these substances burn or decay, the carbon in them unites with oxygen and forms carbon dioxide.

The food which we eat is either animal or vegetable, and it is made ready for bodily use by a slow process of burning within the body; carbon dioxide accompanies this bodily burning of food just as it accompanies the fires with which we are more familiar. The carbon dioxide thus produced within the body escapes into the atmosphere with the breath.

We see that the source of carbon dioxide is practically inexhaustible, coming as it does from every stove, furnace, and candle, and further with every breath of a living organism.

46. Danger of Carbon Dioxide. When carbon dioxide occurs in large quantities, it is dangerous to health, because it interferes with normal breathing, lessening the escape of waste matter through the breath and preventing the access to the lungs of the oxygen necessary for life. Carbon dioxide is not poisonous, but it cuts off the supply of oxygen, just as water cuts it off from a drowning man.

Since every man, woman, and child constantly breathes forth carbon dioxide, the danger in overcrowded rooms is great, and proper ventilation is of vital importance.

47. Ventilation. In estimating the quantity of air necessary to keep a room well aired, we must take into account the number of lights (electric lights do not count) to be used, and the number of people to occupy the room. The average house should provide at the *minimum* 600 cubic feet of space for each person, and in addition, arrangements for allowing at least 300 cubic feet of fresh air per person to enter every hour.

In houses which have not a ventilating system, the air should be kept fresh by intelligent action in the opening of doors and windows; and since relatively few houses are equipped with a satisfactory system, the following suggestions relative to intelligent ventilation are offered.

1. Avoid drafts in ventilation.

2. Ventilate on the sheltered side of the house. If the wind is blowing from the north, open south windows.

CHAPTER IV

48. What Becomes of the Carbon Dioxide. When we reflect that carbon dioxide is constantly being supplied to the atmosphere and that it is injurious to health, the question naturally arises as to how the air remains free enough of the gas to support life. This is largely because carbon dioxide is an essential food of plants. Through their leaves plants absorb it from the atmosphere, and by a wonderful process break it up into its component parts, oxygen and carbon. They reject the oxygen, which passes back to the air, but they retain the carbon, which becomes a part of the plant structure. Plants thus serve to keep the atmosphere free from an excess of carbon dioxide and, in addition, furnish oxygen to the atmosphere.

[Illustration: FIG. 24.--Making carbon dioxide from marble and hydrochloric acid.]

49. How to Obtain Carbon Dioxide. There are several ways in which carbon dioxide can be produced commercially, but for laboratory use the simplest is to mix in a test tube powdered marble, or chalk, and hydrochloric acid, and to collect the effervescing gas as shown in Figure 24. The substance which remains in the test tube after the gas has passed off is a solution of a salt and water. From a mixture of hydrochloric acid (HCl) and marble are obtained a salt, water, and carbon dioxide, the desired gas.

50. A Commercial Use of Carbon Dioxide. If a lighted splinter is thrust into a test tube containing carbon dioxide, it is promptly extinguished, because carbon dioxide cannot support combustion; if a stream of carbon dioxide and water falls upon a fire, it acts like a blanket, covering the flames and extinguishing them. The value of a fire extinguisher depends upon the amount of carbon dioxide and water which it can furnish. A fire extinguisher is a metal case containing a solution of bicarbonate of soda, and a glass vessel full of strong sulphuric acid. As long as the extinguisher is in an upright position, these substances are kept separate, but when the extinguisher is inverted, the acid escapes from the bottle, and mixes with the soda solution. The mingling liquids interact and liberate carbon dioxide. A part of the gas thus liberated dissolves in the water of the soda solution and escapes from the tube with the outflowing liquid, while a portion remains undissolved and escapes as a stream of gas. The fire

CHAPTER IV   41

extinguisher is therefore the source of a liquid containing the fire-extinguishing substance and further the source of a stream of carbon dioxide gas.

[Illustration: FIG. 25.--Inside view of a fire extinguisher.]

51. Carbon. Although carbon dioxide is very injurious to health, both of the substances of which it is composed are necessary to life. We ourselves, our bones and flesh in particular, are partly carbon, and every animal, no matter how small or insignificant, contains some carbon; while the plants around us, the trees, the grass, the flowers, contain a by no means meager quantity of carbon.

Carbon plays an important and varied role in our life, and, in some one of its many forms, enters into the composition of most of the substances which are of service and value to man. The food we eat, the clothes we wear, the wood and coal we burn, the marble we employ in building, the indispensable soap, and the ornamental diamond, all contain carbon in some form.

52. Charcoal. One of the most valuable forms of carbon is charcoal; valuable not in the sense that it costs hundreds of dollars, but in the more vital sense, that its use adds to the cleanliness, comfort, and health of man.

The foul, bad-smelling gases which arise from sewers can be prevented from escaping and passing to streets and buildings by placing charcoal filters at the sewer exits. Charcoal is porous and absorbs foul gases, and thus keeps the region surrounding sewers sweet and clean and free of odor. Good housekeepers drop small bits of charcoal into vases of flowers to prevent discoloration of the water and the odor of decaying stems.

If impure water filters through charcoal, it emerges pure, having left its impurities in the pores of the charcoal. Practically all household filters of drinking water are made of charcoal. But such a device may be a source of disease instead of a prevention of disease, unless the filter is regularly cleaned or renewed. This is because the pores soon become clogged with the impurities, and unless they are cleaned, the water

CHAPTER IV

which flows through the filter passes through a bed of impurities and becomes contaminated rather than purified. Frequent cleansing or renewal of the filter removes this difficulty.

Commercially, charcoal is used on a large scale in the refining of sugars, sirups, and oils. Sugar, whether it comes from the maple tree, or the sugar cane, or the beet, is dark colored. It is whitened by passage through filters of finely pulverized charcoal. Cider and vinegar are likewise cleared by passage through charcoal.

The value of carbon, in the form of charcoal, as a purifier is very great, whether we consider it a deodorizer, as in the case of the sewage, or a decolorizer, as in the case of the refineries, or whether we consider the service it has rendered man in the elimination of danger from drinking water.

53. How Charcoal is Made. Charcoal may be made by heating wood in an oven to which air does not have free access. The absence of air prevents ordinary combustion, nevertheless the intense heat affects the wood and changes it into new substances, one of which is charcoal.

The wood which smolders on the hearth and in the stove is charcoal in the making. Formerly wood was piled in heaps, covered with sod or sand to prevent access of oxygen, and then was set fire to; the smoldering wood, cut off from an adequate supply of air, was slowly transformed into charcoal. Scattered over the country one still finds isolated charcoal kilns, crude earthen receptacles, in which wood thus deprived of air was allowed to smolder and form charcoal. To-day charcoal is made commercially by piling wood on steel cars and then pushing the cars into strong walled chambers. The chambers are closed to prevent access of air, and heated to a high temperature. The intense heat transforms the wood into charcoal in a few hours. A student can make in the laboratory sufficient charcoal for art lessons by heating in an earthen vessel wood buried in sand. The process will be slow, however, because the heat furnished by a Bunsen burner is not great, and the wood is transformed slowly.

A form of charcoal known as animal charcoal, or bone black, is obtained from the charred remains of animals rather than plants, and may be prepared by burning bones and animal refuse as in the case of the wood.

Destructive Distillation. When wood is burned without sufficient air, it is changed into soft brittle charcoal, which is very different from wood. It weighs only one fourth as much as the original wood. It is evident that much matter must leave the wood during the process of charcoal making. We can prove this by putting some dry shavings in a strong test tube fitted with a delivery tube. When the wood is heated a gas passes off which we may collect and burn. Other substances also come off in gaseous form, but they condense in the water. Among these are wood alcohol, wood tar, and acetic acid. In the older method of charcoal making all these products were lost. Can you give any uses of these substances?

54. Matter and Energy. When wood is burned, a small pile of ashes is left, and we think of the bulk of the wood as destroyed. It is true we have less matter that is available for use or that is visible to sight, but, nevertheless, no matter has been destroyed. The matter of which the wood is composed has merely changed its character, some of it is in the condition of ashes, and some in the condition of invisible gases, such as carbon dioxide, but none of it has been destroyed. It is a principle of science that matter can neither be destroyed nor created; it can only be changed, or transformed, and it is our business to see that we do not heedlessly transform it into substances which are valueless to us and our descendants; as, for example, when our magnificent forests are recklessly wasted. The smoke, gases, and ashes left in the path of a raging forest fire are no compensation to us for the valuable timber destroyed. The sum total of matter has not been changed, but the amount of matter which man can use has been greatly lessened.

The principle just stated embodies one of the fundamental laws of science, called the law of *conservation of matter*.

A similar law holds for energy as well. We can transform electric energy into the motion of trolley cars, or we can make use of the energy of streams to turn the wheels of our mills, but in all these cases we are transforming, not creating, energy.

When a ball is fired from a rifle, most of the energy of the gunpowder is utilized in motion, but some is dissipated in producing a flash and a report, and in heat. The energy of the gunpowder has been scattered, but the sum of the various forms of energy is equal to the energy originally stored away in the powder. The better the gun is, the less will be the energy dissipated in smoke and heat and noise.

CHAPTER V

FOOD

55. The Body as a Machine. Wholesome food and fresh air are necessary for a healthy body. Many housewives, through ignorance, supply to their hard-working husbands and their growing sons and daughters food which satisfies the appetite, but which does not give to the body the elements needed for daily work and growth. Some foods, such as lettuce, cucumbers, and watermelons, make proper and satisfactory changes in diet, but are not strength giving. Other foods, like peas and beans, not only satisfy the appetite, but supply to the body abundant nourishment. Many immigrants live cheaply and well with beans and bread as their main diet.

It is of vital importance that the relative value of different foods as heat producers be known definitely; and just as the yard measures length and the pound measures weight the calorie is used to measure the amount of heat which a food is capable of furnishing to the body. Our bodies are human machines, and, like all other machines, require fuel for their maintenance. The fuel supplied to an engine is not all available for pulling the cars; a large portion of the fuel is lost in smoke, and another portion is wasted as ashes. So it is with the fuel that runs the body. The food we eat is not all available for nourishment, much of it being as useless to us as are smoke and ashes to an engine. The best foods are those which do the most for us with the least possible waste.

CHAPTER V 45

56. Fuel Value. By fuel value is meant the capacity foods have for yielding heat to the body. The fuel value of the foods we eat daily is so important a factor in life that physicians, dietitians, nurses, and those having the care of institutional cooking acquaint themselves with the relative fuel values of practically all of the important food substances. The life or death of a patient may be determined by the patient's diet, and the working and earning capacity of a father depends largely upon his prosaic three meals. An ounce of fat, whether it is the fat of meat or the fat of olive oil or the fat of any other food, produces in the body two and a quarter times as much heat as an ounce of starch. Of the vegetables, beans provide the greatest nourishment at the least cost, and to a large extent may be substituted for meat. It is not uncommon to find an outdoor laborer consuming one pound of beans per day, and taking meat only on "high days and holidays."

[Illustration: FIG. 26.--The bomb calorimeter from which the fuel value of food can be estimated.]

The fuel value of a food is determined by means of the _bomb calorimeter (Fig. 26). The food substance is put into a chamber A_ and ignited, and the heat of the burning substance raises the temperature of the water in the surrounding vessel. If 1000 grams of water are in the vessel, and the temperature of the water is raised 2° C., the number of calories produced by the substance would be 2000, and the fuel value would be 2000 calories.[A] From this the fuel value of one quart or one pound of the substance can be determined, and the food substance will be said to furnish the body with that number of heat units, providing all of the pound of food were properly digested.

[Footnote A: As applied to food, the calorie is greater than that used in the ordinary laboratory work, being the amount of heat necessary to raise the temperature of 1000 grams of water 1° C., rather than 1 gram 1° C.]

TABLE SHOWING THE NUMBER OF CALORIES FURNISHED BY ONE POUND OF VARIOUS FOODS
--- |FOOD |CALORIES|FOOD |CALORIES| --- |Leg of lean mutton | 790|Carrots | 210| ---

CHAPTER V
46

Rib of beef	1150	Lettuce	90
Shad	380	Onion	225
Chicken	505	Cucumber	80
Apples	290	Almonds	3030
Bananas	460	Walnuts	3306
Prunes	370	Peanuts	2560
Watermelons	140	Oatmeal	4673
Lima beans	570	Rolled wheat	4175
Beets	215	Macaroni	1665

57. Varied Diet. The human body is a much more varied and complex machine than any ever devised by man; personal peculiarities, as well as fuel values, influence very largely the diet of an individual. Strawberries are excluded from some diets because of a rash which is produced on the skin, pork is excluded from other diets for a like reason; cauliflower is absolutely indigestible to some and is readily digested by others. From practically every diet some foods must be excluded, no matter what the fuel value of the substance may be.

Then, too, there are more uses for food than the production of heat. Teeth and bones and nails need a constant supply of mineral matter, and mineral matter is frequently found in greatest abundance in foods of low fuel value, such as lettuce, watercress, etc., though practically all foods yield at least a small mineral constituent. When fuel values alone are considered, fruits have a low value, but because of the flavor they impart to other foods, and because of the healthful influence they exercise in digestion, they cannot be excluded from the diet.

Care should be constantly exercised to provide substantial foods of high fuel value. But the nutritive foods should be wisely supplemented by such foods as fruits, whose real value is one of indirect rather then direct service.

58. Our Bodies. Somewhat as a house is composed of a group of bricks, or a sand heap of grains of sand, the human body is composed of small divisions called cells. Ordinarily we cannot see these cells because of their minuteness, but if we examine a piece of skin, or a

CHAPTER V 47

hair of the head, or a tiny sliver of bone under the microscope, we see that each of these is composed of a group of different cells. A merchant, watchful about the fineness of the wool which he is purchasing, counts with his lens the number of threads to the inch; a physician, when he wishes, can, with the aid of the microscope, examine the cells in a muscle, or in a piece of fat, or in a nerve fiber. Not only is the human body composed of cells, but so also are the bodies of all animals from the tiny gnat which annoys us, and the fly which buzzes around us, to the mammoth creatures of the tropics. These cells do the work of the body, the bone cells build up the skeleton, the nail cells form the finger and toe nails, the lung cells take care of breathing, the muscle cells control motion, and the brain cells are responsible for thought.

59. Why we eat so Much. The cells of the body are constantly, day by day, minute by minute, breaking down and needing repair, are constantly requiring replacement by new cells, and, in the case of the child, are continually increasing in number. The repair of an ordinary machine, an engine, for example, is made at the expense of money, but the repair and replacement of our human cell machinery are accomplished at the expense of food. More than one third of all the food we eat goes to maintain the body cells, and to keep them in good order. It is for this reason that we consume a large quantity of food. If all the food we eat were utilized for energy, the housewife could cook less, and the housefather could save money on grocer's and butcher's bills. If you put a ton of coal in an engine, its available energy is used to run the engine, but if the engine were like the human body, one third of the ton would be used up by the engine in keeping walls, shafts, wheels, belts, etc., in order, and only two thirds would go towards running the engine. When an engine is not working, fuel is not consumed, but the body requires food for mere existence, regardless of whether it does active work or not. When we work, the cells break down more quickly, and the repair is greater than when we are at rest, and hence there is need of a larger amount of food; but whether we work or not, food is necessary.

60. The Different Foods. The body is very exacting in its demands, requiring certain definite foods for the formation and maintenance of its cells, and other foods, equally definite, but of different character, for

heat; our diet therefore must contain foods of high fuel value, and likewise foods of cell-forming power.

Although the foods which we eat are of widely different character, such as fruits, vegetables, cereals, oils, meats, eggs, milk, cheese, etc., they can be put into three great classes: the carbohydrates, the fats, and the proteids.

61. The Carbohydrates. Corn, wheat, rye, in fact all cereals and grains, potatoes, and most vegetables are rich in carbohydrates; as are also sugar, molasses, honey, and maple sirup. The foods of the first group are valuable because of the starch they contain; for example, corn starch, wheat starch, potato starch. The substances of the second group are valuable because of the sugar they contain; sugar contains the maximum amount of carbohydrate. In the sirups there is a considerable quantity of sugar, while in some fruits it is present in more or less dilute form. Sweet peaches, apples, grapes, contain a moderate amount of sugar; watermelons, pears, etc., contain less. Most of our carbohydrates are of plant origin, being found in vegetables, fruits, cereals, and sirups.

Carbohydrates, whether of the starch group or the sugar group, are composed chiefly of three elements: carbon, hydrogen, and oxygen; they are therefore combustible, and are great energy producers. On the other hand, they are worthless for cell growth and repair, and if we limited our diet to carbohydrates, we should be like a man who had fuel but no engine capable of using it.

62. The Fats. The best-known fats are butter, lard, olive oil, and the fats of meats, cheese, and chocolate. When we test fats for fuel values by means of a calorimeter (Fig. 26), we find that they yield twice as much heat as the carbohydrates, but that they burn out more quickly. Dwellers in cold climates must constantly eat large quantities of fatty foods if they are to keep their bodies warm and survive the extreme cold. Cod liver oil is an excellent food medicine, and if taken in winter serves to warm the body and to protect it against the rigors of cold weather. The average person avoids fatty foods in summer, knowing from experience that rich foods make him warm and uncomfortable. The harder we work and the colder the weather, the more food of that

CHAPTER V   49

kind do we require; it is said that a lumberman doing heavy out-of-door work in cold climates needs three times as much food as a city clerk. Most of our fats, like lard and butter, are of animal origin; some of them, however, like olive oil, peanut butter, and coconut oil, are of plant origin.

[Illustration: FIG. 27.--*a* is the amount of fat necessary to make one calorie; *b* is the amount of sugar or proteid necessary to make one calorie.]

63. The Proteids. The proteids are the building foods, furnishing muscle, bone, skin cells, etc., and supplying blood and other bodily fluids. The best-known proteids are white of egg, curd of milk, and lean of fish and meat; peas and beans have an abundant supply of this substance, and nuts are rich in it. Most of our proteids are of animal origin, but some protein material is also found in the vegetable world. This class of foods contains carbon, oxygen, and hydrogen, and in addition, two substances not found in carbohydrates or fats--namely, sulphur and nitrogen. Proteids always contain nitrogen, and hence they are frequently spoken of as nitrogenous foods. Since the proteids contain all the elements found in the two other classes of foods, they are able to contribute, if necessary, to the store of bodily energy; but their main function is upbuilding, and the diet should be chosen so that the proteids do not have a double task.

For an average man four ounces of dry protein matter daily will suffice to keep the body cells in normal condition.

It has been estimated that 300,000,000 blood cells alone need daily repair or renewal. When we consider that the blood is but one part of the body, and that all organs and fluids have corresponding requirements, we realize how vast is the work to be done by the food which we eat.

64. Mistakes in Buying. The body demands a daily ration of the three classes of food stuffs, but it is for us to determine from what meats, vegetables, fruits, cereals, etc., this supply shall be obtained (Figs. 28 and 29).

CHAPTER V

[Illustration: FIG. 28.--Table of food values.]

[Illustration: FIG. 29.--Diagram showing the difference in the cost of three foods which give about the same amount of nutrition each.]

Generally speaking, meats are the most expensive foods we can purchase, and hence should be bought seldom and in small quantities. Their place can be taken by beans, peas, potatoes, etc., and at less than a quarter of the cost. The average American family eats meat three times a day, while the average family of the more conservative and older countries rarely eats meat more than once a day. The following tables indicate the financial loss arising from an unwise selection of foods:--

FOOD CONSUMED--ONE WEEK

FAMILY No. 1		FAMILY No. 2	
20 loaves of bread	$1.00	15 lb. flour, bread home made (skim milk used)	$.45
10 to 12 lb. loin steak or meat of similar cost	2.00	Yeast, shortening, and skim milk	.10
20 to 25 lb. rib roast or similar meat	4.40	4 lb. high-priced cereal breakfast food, 20¢	.80
10 lb. steak (round, Hamburger and some loin)	1.50	5 lb. cheese, 16¢	.80
10 lb. other meats, boiling pieces, rump roast, etc.	1.00	5 lb. oatmeal (bulk)	.15
Cake and pastry purchased	3.00	Mushrooms	.75
8 lb. butter, 30¢	2.40	5 lb. beans	.25
Tea, coffee, spices, etc.	.75	Home-made cake and pastry	1.00
Celery	1.00	6 lb. butter, 30¢	1.80
Oranges	2.00	3 lb. home-made shortening	.25
Potatoes	.25	Tea, coffee, and spices	.40
Miscellaneous canned goods	2.00	Milk	.50
Miscellaneous foods	2.00	Apples	.50
3 doz. eggs	.60	Prunes	.25
Milk	1.00	Potatoes	.25
		Miscellaneous foods	1.00
		3 doz. eggs	.60
	$23.45		$11.30

"The tables show that one family spends over twice as much in the purchase of foods as the other family, and yet the one whose food costs the less actually secures the larger amount of nutritive material

CHAPTER VI
51

and is better fed than the family where more money is expended."--From *Human Foods*, Snyder.

The Source of the Different Foods. All of our food comes from either the plant world or the animal world. Broadly speaking, plants furnish the carbohydrates, that is, starch and sugar; animals furnish the fats and proteids. But although vegetable foods yield carbohydrates mainly, some of them, like beans and peas, contain large quantities of protein and can be substituted for meat without disadvantage to the body. Other plant products, such as nuts, have fat as their most abundant food constituent. The peanut, for example, contains 43% of fat, 30% of proteids, and only 17% of carbohydrates; the Brazil nut has 65% of fat, 17% of proteids, and only 9% of carbohydrates. Nuts make a good meat substitute, and since they contain a fair amount of carbohydrates besides the fats and proteins, they supply all of the essential food constituents and form a well-balanced food.

CHAPTER VI

WATER

65. Destructive Action of Water. The action of water in stream and sea, in springs and wells, is evident to all; but the activity of ground water--that is, rain water which sinks into the soil and remains there--is little known in general. The real activity of ground water is due to its great solvent power; every time we put sugar into tea or soap into water we are using water as a solvent. When rain falls, it dissolves substances floating in the atmosphere, and when it sinks into the ground and becomes ground water, it dissolves material out of the rock which it encounters (Fig. 30). We know that water contains some mineral matter, because kettles in which water is boiled acquire in a short time a crust or coating on the inside. This crust is due to the accumulation in the kettle of mineral matter which was in solution in the water, but which was left behind when the water evaporated. (See Section 25.)

[Illustration: FIG. 30.--Showing how caves and holes are formed by the solvent action of water.]

The amount of dissolved mineral matter present in some wells and springs is surprisingly great; the famous springs of Bath, England, contain so much mineral matter in solution, that a column 9 feet in diameter and 140 feet high could be built out of the mineral matter contained in the water consumed yearly by the townspeople.

[Illustration: FIG. 31.--The work of water as a solvent.]

Rocks and minerals are not all equally soluble in water; some are so little soluble that it is years before any change becomes apparent, and the substances are said to be insoluble, yet in reality they are slowly dissolving. Other rocks, like limestone, are so readily soluble in water that from the small pores and cavities eaten out by the water, there may develop in long centuries, caves and caverns (Fig. 30). Most rock, like granite, contains several substances, some of which are readily soluble and others of which are not readily soluble; in such rocks a peculiar appearance is presented, due to the rapid disappearance of the soluble substance, and the persistence of the more resistant substance (Fig. 31).

We see that the solvent power of water is constantly causing changes, dissolving some mineral substances, and leaving others practically untouched; eating out crevices of various shapes and sizes, and by gradual solution through unnumbered years enlarging these crevices into wonderful caves, such as the Mammoth Cave of Kentucky.

66. Constructive Action of Water. Water does not always act as a destructive agent; what it breaks down in one place it builds up in another. It does this by means of precipitation. Water dissolves salt, and also dissolves lead nitrate, but if a salt solution is mixed with a lead nitrate solution, a solid white substance is formed in the water (Fig. 32). This formation of a solid substance from the mingling of two liquids is called precipitation; such a process occurs daily in the rocks beneath the surface of the earth. (See Laboratory Manual.)

[Illustration: FIG. 32.--From the mingling of two liquids a solid is sometimes formed.]

Suppose water from different sources enters a crack in a rock, bringing different substances in solution; then the mingling of the waters may cause precipitation, and the solid thus formed will be deposited in the crack and fill it up. Hence, while ground water tends to make rock porous and weak by dissolving out of it large quantities of mineral matter, it also tends under other conditions to make it more compact because it deposits in cracks, crevices, and pores the mineral matter precipitated from solution.

These two forces are constantly at work; in some places the destructive action is more prominent, in other places the constructive action; but always the result is to change the character of the original substance. When the mineral matter precipitated from the solutions is deposited in cracks, *veins* are formed (Fig. 33), which may consist of the ore of different metals, such as gold, silver, copper, lead, etc. Man is almost entirely dependent upon these veins for the supply of metal needed in the various industries, because in the original condition of the rocks, the metallic substances are so scattered that they cannot be profitably extracted.

[Illustration: FIG. 33.--Mineral matter precipitated from solution is deposited in crevices and forms veins.]

Naturally, the veins themselves are not composed of one substance alone, because several different precipitates may be formed. But there is a decided grouping of valuable metals, and these can then be readily separated by means of electricity.

67. Streams. Streams usually carry mud and sand along with them; this is particularly well seen after a storm when rivers and brooks are muddy. The puddles which collect at the foot of a hill after a storm are muddy because of the particles of soil gathered by the water as it runs down the hill. The particles are not dissolved in the water, but are held there in suspension, as we call it technically. The river made muddy after a storm by suspended particles usually becomes clear and transparent after it has traveled onward for miles, because, as it

CHAPTER VI 54

travels, the particles drop to the bottom and are deposited there. Hence, materials suspended in the water are borne along and deposited at various places (Fig. 34). The amount of deposition by large rivers is so great that in some places channels fill up and must be dredged annually, and vessels are sometimes caught in the deposit and have to be towed away.

Running water in the form of streams and rivers, by carrying sand particles, stones, and rocks from high slopes and depositing them at lower levels, wears away land at one place and builds it up at another, and never ceases in its work of changing the nature of the earth's surface (Fig. 35).

[Illustration: FIG. 34.--Deposit left by running water.]

[Illustration: FIG. 35.--Water by its action constantly changes the character of the land.]

68. Relation of Water to Human Life. Water is one of the most essential of food materials, and whether we drink much or little water, we nevertheless get a great deal of it. The larger part of many of our foods is composed of water; more than half of the weight of the meat we eat is made up of water; and vegetables are often more than nine tenths water. (See Laboratory Manual.) Asparagus and tomatoes have over 90 per cent. of water, and most fruits are more than three fourths water; even bread, which contains as little water as any of our common foods, is about one third water (Fig. 36).

[Illustration: FIG. 36.--Diagram of the composition of a loaf of bread and of a potato: 1. ash; 2, food; 3, water.]

Without water, solid food material, although present in the body, would not be in a condition suitable for bodily use. An abundant supply of water enables the food to be dissolved or suspended in it, and in solution the food material is easily distributed to all parts of the body.

Further, water assists in the removal of the daily bodily wastes, and thus rids the system of foul and poisonous substances.

CHAPTER VI

The human body itself consists largely of water; indeed, about two thirds of our own weight is water. The constant replenishing of this large quantity is necessary to life, and a considerable amount of the necessary supply is furnished by foods, particularly the fruits and vegetables.

But while the supply furnished by the daily food is considerable, it is by no means sufficient, and should be supplemented by good drinking water.

69. Water and its Dangers. Our drinking water comes from far and near, and as it moves from place to place, it carries with it in solution or suspension anything which it can find, whether it be animal, vegetable, or mineral matter. The power of water to gather up matter is so great that the average drinking water contains 20 to 90 grains of solid matter per gallon; that is, if a gallon of ordinary drinking water is left to evaporate, a residue of 20 to 90 grains will be left. (See Laboratory Manual.) As water runs down a hill slope (Fig. 37), it carries with it the filth gathered from acres of land; carries with it the refuse of stable, barn, and kitchen; and too often this impure surface water joins the streams which supply our cities. Lakes and rivers which furnish drinking water should be carefully protected from surface draining; that is, from water which has flowed over the land and has thus accumulated the waste of pasture and stable and, it may be, of dumping ground.

[Illustration: FIG. 37.--As water flows over the land, it gathers filth and disease germs.]

It is not necessary that water should be absolutely free from all foreign substances in order to be safe for daily use in drinking; a limited amount of mineral matter is not injurious and may sometimes be really beneficial. It is the presence of animal and vegetable matter that causes real danger, and it is known that typhoid fever is due largely to such impurities present in the drinking water.

70. Methods of Purification. Water is improved by any of the following methods:--

(*a*) *Boiling.* The heat of boiling destroys animal and vegetable germs. Hence water that has been boiled a few minutes is safe to use. This is the most practical method of purification in the home, and is very efficient. The boiled water should be kept in clean, corked bottles; otherwise foreign substances from the atmosphere reënter the water, and the advantage gained from boiling is lost.

(*b*) *Distillation.* By this method pure water is obtained, but this method of purification cannot be used conveniently in the home (Section 25).

(*c*) *Filtration.* In filtration, the water is forced through porcelain or other porous substances which allow the passage of water, but which hold back the minute foreign particles suspended in the water. (See Laboratory Manual.) The filters used in ordinary dwellings are of stone, asbestos, or charcoal. They are often valueless, because they soon become choked and cannot be properly cleaned.

The filtration plants owned and operated by large cities are usually safe; there is careful supervision of the filters, and frequent and effective cleanings are made. In many cities the filtration system is so good that private care of the water supply is unnecessary.

71. The Source of Water. In the beginning, the earth was stored with water just as it was with metal, rock, etc. Some of the water gradually took the form of rivers, lakes, streams, and wells, as now, and it is this original supply of water which furnishes us all that we have to-day. We quarry to obtain stone and marble for building, and we fashion the earth's treasures into forms of our own, but we cannot create these things. We bore into the ground and drill wells in order to obtain water from hidden sources; we utilize rapidly flowing streams to drive the wheels of commerce, but the total amount of water remains practically unchanged.

The water which flows on the earth is constantly changing its form; the heat of the sun causes it to evaporate, or to become vapor, and to mingle with the atmosphere. In time, the vapor cools, condenses, and falls as snow or rain; the water which is thus returned to the earth feeds our rivers, lakes, springs, and wells, and these in turn supply water to man. When water falls upon a field, it soaks into the ground,

CHAPTER VI 57

or collects in puddles which slowly evaporate, or it runs off and drains into small streams or into rivers. That which soaks into the ground is the most valuable because it remains on the earth longest and is the purest.

[Illustration: FIG. 38.--How springs are formed. *A*, porous layer; *B*, non-porous layer; *C*, spring.]

Water which soaks into the ground moves slowly downward and after a longer or shorter journey, meets with a non-porous layer of rock through which it cannot pass, and which effectually hinders its downward passage. In such regions, there is an accumulation of water, and a well dug there would have an abundant supply of water. The non-porous layer is rarely level, and hence the water whose vertical path is obstructed does not "back up" on the soil, but flows down hill parallel with the obstructing non-porous layer, and in some distant region makes an outlet for itself, forming a spring (Fig. 38). The streams originating in the springs flow through the land and eventually join larger streams or rivers; from the surface of streams and rivers evaporation occurs, the water once more becomes vapor and passes into the atmosphere, where it is condensed and again falls to the earth.

Water which has filtered through many feet of earth is far purer and safer than that which fell directly into the rivers, or which ran off from the land and joined the surface streams without passing through the soil.

72. The Composition of Water. Water was long thought to be a simple substance, but toward the end of the eighteenth century it was found to consist of two quite different substances, oxygen (O) and hydrogen (H.)

[Illustration: FIG. 39.--The decomposition of water.]

If we send an electric current through water (acidulated to make it a good conductor), as shown in Figure 39, we see bubbles of gas rising from the end of the wire by which the current enters the water, and other bubbles of gas rising from the end of the wire by which the current leaves the water. These gases have evidently come from the water and are the substances of which it is composed, because the water begins to disappear as the gases are formed. If we place over each end of the wire an inverted jar filled with water, the gases are easily collected. The first thing we notice is that there is always twice as much of one gas as of the other; that is, water is composed of two substances, one of which is always present in twice as large quantities as the other.

73. The Composition of Water. On testing the gases into which water is broken up by an electric current, we find them to be quite different. One proves to be oxygen, a substance with which we are already familiar. The other gas, hydrogen, is new to us and is interesting as being the lightest known substance, being even "lighter than a feather."

An important fact about hydrogen is that in burning it gives as much heat as five times its weight of coal. Its flame is blue and almost invisible by daylight, but intensely hot. If fine platinum wire is placed in an ordinary gas flame, it does not melt, but if placed in a flame of burning hydrogen, it melts very quickly.

74. How to prepare Hydrogen. There are many different methods of preparing hydrogen, but the easiest laboratory method is to pour sulphuric acid, or hydrochloric acid, on zinc shavings and to collect in a bottle the gas which is given off. This gas proves to be colorless, tasteless, and odorless. (See Laboratory Manual.)

CHAPTER VII

AIR

75. The Instability of the Air. We are usually not conscious of the air around us, but sometimes we realize that the air is heavy, while at other times we feel the bracing effect of the atmosphere. We live in an ocean of air as truly as fish inhabit an ocean of water. If you have ever been at the seashore you know that the ocean is never still for a second; sometimes the waves surge back and forth in angry fury, at other times the waves glide gently in to the shore and the surface is as

CHAPTER VII 59

smooth as glass; but we know that there is perpetual motion of the water even when the ocean is in its gentlest moods. Generally our atmosphere is quiet, and we are utterly unconscious of it; at other times we are painfully aware of it, because of its furious winds. Then again we are oppressed by it because of the vast quantity of vapor which it holds in the form of fog, or mist. The atmosphere around us is as restless and varying as is the water of the sea. The air at the top of a high tower is very different from the air at the base of the tower. Not only does the atmosphere vary greatly at different altitudes, but it varies at the same place from time to time, at one period being heavy and raw, at another being fresh and invigorating.

Winds, temperature, and humidity all have a share in determining atmospheric conditions, and no one of these plays a small part.

76. The Character of the Air. The atmosphere which envelops us at all times extends more than fifty miles above us, its height being far greater than the greatest depths of the sea. This atmosphere varies from place to place; at the sea level it is heavy, on the mountain top less heavy, and far above the earth it is so light that it does not contain enough oxygen to permit man to live. Figure 40 illustrates by a pile of pillows how the pressure of the air varies from level to level.

[Illustration: FIG. 40.--To illustrate the decrease in pressure with height.]

Sea level is a low portion of the earth's surface, hence at sea level there is a high column of air, and a heavy air pressure. As one passes from sea level to mountain top a gradual but steady decrease in the height of the air column occurs, and hence a gradual but definite lessening of the air pressure.

[Illustration: FIG. 41.--The water in the tube is at the same level as that in the glass.]

77. Air Pressure. If an empty tube (Fig. 41) is placed upright in water, the water will not rise in the tube, but if the tube is put in water and the air is then drawn out of the tube by the mouth, the water will rise in the tube (Fig. 42). This is what happens when we take lemonade through

CHAPTER VII 60

a straw. When the air is withdrawn from the straw by the mouth, the pressure within the straw is reduced, and the liquid is forced up the straw by the air pressure on the surface of the liquid in the glass. Even the ancient Greeks and Romans knew that water would rise in a tube when the pressure within the tube was reduced, and hence they tried to obtain water from wells in this fashion, but the water could never be raised higher than 34 feet. Let us see why water could rise 34 feet and no more. If an empty pipe is placed in a cistern of water, the water in the pipe does not rise above the level of the water in the cistern. If, however, the pressure in the tube is removed, the water in the tube will rise to a height of 34 feet approximately. If now the air pressure in the tube is restored, the water in the tube sinks again to the level of that in the cistern. The air pressing on the liquid in the cistern tends to push some liquid up the tube, but the air pressing on the water in the tube pushes downwards, and tends to keep the liquid from rising, and these two pressures balance each other. When, however, the pressure within the tube is reduced, the liquid rises because of the unbalanced pressure which acts on the water in the cistern.

[Illustration: FIG. 42.--Water rises in the tube when the air is withdrawn.]

[Illustration: FIG. 43.--The air supports a column of mercury 30 inches high.]

The column of water which can be raised this way is approximately 34 feet, sometimes a trifle more, sometimes a trifle less. If water were twice as heavy, just half as high a column could be supported by the atmosphere. Mercury is about thirteen times as heavy as water and, therefore, the column of mercury supported by the atmosphere is about one thirteenth as high as the column of water supported by the atmosphere. This can easily be demonstrated. Fill a glass tube about a yard long with mercury, close the open end with a finger, and quickly insert the end of the inverted tube in a dish of mercury (Fig. 43). When the finger is removed, the mercury falls somewhat, leaving an empty space in the top of the tube. If we measure the column in the tube, we find its height is about one thirteenth of 34 feet or 30 inches, exactly what we should expect. Since there is no air pressure within the tube, the atmospheric pressure on the mercury in the dish is balanced solely by the mercury within the tube, that is, by a column of mercury 30 inches high. The shortness of the mercury column as compared with that of water makes the mercury more convenient for both experimental and practical purposes. (See Laboratory Manual.)

78. The Barometer. Since the pressure of the air changes from time to time, the height of the mercury will change from day to day, and hour to hour. When the air pressure is heavy, the mercury will tend to be high; when the air pressure is low, the mercury will show a shorter column; and by reading the level of the mercury one can learn the pressure of the atmosphere. If a glass tube and dish of mercury are attached to a board and the dish of mercury is inclosed in a case for protection from moisture and dirt, and further if a scale of inches or centimeters is made on the upper portion of the board, we have a mercurial barometer (Fig. 44).

[Illustration: FIG. 44.--A simple barometer.]

If the barometer is taken to the mountain top, the column of mercury falls gradually during the ascent, showing that as one ascends, the pressure decreases in agreement with the statement in Section 76. Observations similar to these were made by Torricelli as early as the sixteenth century. Taking a barometric reading consists in measuring the height of the mercury column.

79. A Portable Barometer. The mercury barometer is large and inconvenient to carry from place to place, and a more portable form has been devised, known as the aneroid barometer (Fig. 45). This form of barometer is extremely sensitive; indeed, it is so delicate that it shows the slight difference between the pressure at the table top and the pressure at the floor level, whereas the mercury barometer would indicate only a much greater variation in atmospheric pressure. The aneroid barometers are frequently made no larger than a watch and can be carried conveniently in the pocket, but they get out of order easily and must be frequently readjusted. The aneroid barometer is an air-tight box whose top is made of a thin metallic disk which bends inward or outward according to the pressure of the atmosphere. If the atmospheric pressure increases, the thin disk is pushed slightly inward; if, on the other hand, the atmospheric pressure decreases, the

CHAPTER VII

pressure on the metallic disk decreases and the disk is not pressed so far inward. The motion of the disk is small, and it would be impossible to calculate changes in atmospheric pressure from the motion of the disk, without some mechanical device to make the slight changes in motion perceptible.

[Illustration: FIG. 45.--Aneroid barometer.]

In order to magnify the slight changes in the position of the disk, the thin face is connected with a system of levers, or wheels, which multiplies the changes in motion and communicates them to a pointer which moves around a graduated circular face. In Figure 45 the real barometer is scarcely visible, being securely inclosed in a metal case for protection; the principle, however, can be understood by reference to Figure 46.

[Illustration: FIG. 46.--Principle of the aneroid barometer.]

80. The Weight of the Air. We have seen that the pressure of the atmosphere at any point is due to the weight of the air column which stretches from that point far up into the sky above. This weight varies slightly from time to time and from place to place, but it is equal to about 15 pounds to the square inch as shown by actual measurement. It comes to us as a surprise sometimes that air actually has weight; for example, a mass of 12 cubic feet of air at average pressure weighs 1 pound, and the air in a large assembly hall weighs more than 1 ton.

We are practically never conscious of this really enormous pressure of the atmosphere, which is exerted over every inch of our bodies, because the pressure is exerted equally over the outside and the inside of our bodies; the cells and tissues of our bodies containing gases under atmospheric pressure. If, however, the finger is placed over the open end of a tube and the air is sucked out of the tube by the mouth, the flesh of the finger bulges into the tube because the pressure within the finger is no longer equalized by the usual atmospheric pressure (Fig. 47).

[Illustration: FIG. 47.--The flesh bulges out.]

CHAPTER VII 63

Aëronauts have never ascended much higher than 7 miles; at that height the barometer stands at 7 inches instead of at 30 inches, and the internal pressure in cells and tissues is not balanced by an equal external pressure. The unequalized internal pressure forces the blood to the surface of the body and causes rupture of blood vessels and other physical difficulties.

81. Use of the Barometer. Changes in air pressure are very closely connected with changes in the weather. The barometer does not directly foretell the weather, but a low or falling pressure, accompanied by a simultaneous fall of the mercury, usually precedes foul weather, while a rising pressure, accompanied by a simultaneous rise in the mercury, usually precedes fair weather. The barometer is not an infallible prophet, but it is of great assistance in predicting the general trend of the weather. There are certain changes in the barometer which follow no known laws, and which allow of no safe predictions, but on the other hand, general future conditions for a few days ahead can be fairly accurately determined. Figure 48 shows a barograph or self-registering barometer which automatically registers air pressure.

[Illustration: FIG. 48.--Barograph.]

Seaport towns in particular, but all cities, large or small, and villages too, are on request notified by the United States Weather Bureau ten hours or more in advance, of probable weather conditions, and in this way precautions are taken which annually save millions of dollars and hundreds of lives.

I recollect a summer spent on a New Hampshire farm, and know that an old farmer started his farm hands haying by moonlight at two o'clock in the morning, because the Special Farmer's Weather Forecast of the preceding evening had predicted rain for the following day. His reliance on the weather report was not misplaced, since the storm came with full force at noon. Sailing vessels, yachts, and fishing dories remain within reach of port if the barometer foretells storms.

[Illustration: FIG. 49.--Isotherms.]

82. Isobaric and Isothermal Lines. If a line were drawn through all points on the surface of the earth having an equal barometric pressure at the same time, such a line would be called an isobar. For example, if the height of barometers in different localities is observed at exactly the same time, and if all the cities and towns which have the same pressure are connected by a line, the curved lines will be called isobars. By the aid of these lines the barometric conditions over a large area can be studied. The Weather Bureau at Washington relies greatly on these isobars for statements concerning local and distant weather forecasts, any shift in isobaric lines showing change in atmospheric pressure.

If a line is drawn through all points on the surface of the earth having the same *temperature* at the same instant, such a line is called an isotherm (Fig. 49).

83. Weather Maps. Scattered over the United States are about 125 Government Weather Stations, at each of which three times a day, at the same instant, accurate observations of the weather are made. These observations, which consist of the reading of barometer and thermometer, the determination of the velocity and direction of the wind, the determination of the humidity and of the amount of rain or snow, are telegraphed to the chief weather official at Washington. From the reports of wind storms, excessive rainfall, hot waves, clearing weather, etc., and their rate of travel, the chief officials predict where the storms, etc., will be at a definite future time. In the United States, the *general* movement of weather conditions, as indicated by the barometer, is from west to east, and if a certain weather condition prevails in the west, it is probable that it will advance eastward, although with decided modifications. So many influences modify atmospheric conditions that unfailing predictions are impossible, but the Weather Bureau predictions prove true in about eight cases out of ten.

The reports made out at Washington are telegraphed on request to cities in this country, and are frequently published in the daily papers, along with the forecast of the local office. A careful study of these reports enables one to forecast to some extent the probable weather conditions of the day.

CHAPTER VII

The first impression of a weather map (Fig. 50) with its various lines and signals is apt to be one of confusion, and the temptation comes to abandon the task of finding an underlying plan of the weather. If one will bear in mind a few simple rules, the complexity of the weather map will disappear and a glance at the map will give one information concerning general weather conditions just as a glance at the thermometer in the morning will give some indication of the probable temperature of the day. (See Laboratory Manual.)

[Illustration: FIG. 50. weather Map]

On the weather map solid lines represent isobars and dotted lines represent isotherms. The direction of the wind at any point is indicated by an arrow which flies with the wind; and the state of the weather--clear, partly cloudy, cloudy, rain, snow, etc.--is indicated by symbols.

84. Components of the Air. The best known constituent of the air is oxygen, already familiar to us as the feeder of the fire without and within the body. Almost one fifth of the air which envelops us is made up of the life-giving oxygen. This supply of oxygen in the air is constantly being used up by breathing animals and glowing fires, and unless there were some constant source of additional supply, the quantity of oxygen in the air would soon become insufficient to support animal life. The unfailing constant source of atmospheric oxygen is plant life (Section 48). The leaves of plants absorb carbon dioxide from the air, and break it up into oxygen and carbon. The plant makes use of the carbon but it rejects the oxygen, which passes back into the atmosphere through the pores of the leaves.

Although oxygen constitutes only one fifth of the atmosphere, it is one of the most abundant and widely scattered of all substances. Almost the whole earth, whether it be rich loam, barren clay, or granite boulder, contains oxygen in some form or other; that is, in combination with other substances. But nowhere, except in the air around us, do we find oxygen free and uncombined with other substances.

A less familiar but more abundant constituent of the atmosphere is the nitrogen. Almost four fifths of the air around us is made up of nitrogen.

CHAPTER VIII

66

If the atmosphere were composed of oxygen alone, the merest flicker of a match would set the whole world ablaze. The fact that the oxygen of the air is diluted as it were with so large a proportion of nitrogen, prevents fires from sweeping over the world and destroying everything in their path. Nitrogen does not support combustion, and a burning match placed in a corked bottle goes out as soon as it has used up the oxygen in the bottle. The nitrogen in the bottle, not only does not assist the burning of the match, but it acts as a damper to the burning.

Free nitrogen, like oxygen, is a colorless, odorless gas. It is not poisonous; but one would die if surrounded by nitrogen alone, just as one would die if surrounded by water. The vast supply of nitrogen in the atmosphere would be useless if the smaller amount of oxygen were not present to keep the body alive. Nitrogen is so important a factor in daily life that an entire chapter will be devoted to it later.

Another constituent of the air with which we are familiar is carbon dioxide. In pure air, carbon dioxide is present in very small proportion, being continually taken from the air by plants in the manufacture of their food.

Various other substances are present in the air in very minute proportions, but of all the substances in the air, oxygen, nitrogen, and carbon dioxide are the most important.

CHAPTER VIII

GENERAL PROPERTIES OF GASES

85. Bicycle Tires. We know very well that we cannot put more than a certain amount of water in a tube, but we know equally well that the amount of air which can be pumped into a bicycle or automobile tire depends largely upon our muscular energy. A gallon of water remains a gallon of water and requires a perfectly definite amount of space, but air can be compressed and compressed, and made to occupy less and less space. While it is true that air is easily compressed, it is also true that air is elastic and capable of very rapid and easy expansion. If a

puncture occurs in a tire, the compressed air escapes very quickly; that is, the compressed air within the tube has taken the first opportunity offered for expansion.

[Illustration: FIG. 51.--By squeezing the bulb, air is forced out of the nozzle.]

The fact that air is elastic has added materially to the comfort of the world. Transportation by bicycles and automobiles has been greatly facilitated by the use of air tires. In many hospitals, air mattresses are used in place of hair, feather, or cotton mattresses, and in this way the bed is kept fresher and cleaner, and can be moved with less danger of discomfort to the patient. Every time we squeeze the bulb of an atomizer, we force compressed or condensed air through the atomizer, and the condensed air pushes the liquid out of the nozzle (Fig. 51). Thus we see that in the necessities and conveniences of life compressed air plays an important part.

86. The Danger of Compression. Air under ordinary atmospheric conditions exerts a pressure of 15 pounds to the square inch. If, now, large quantities of air are compressed into a small space, the pressure exerted becomes correspondingly greater. If too much air is blown into a toy balloon, the balloon bursts because it cannot support the great pressure exerted by the compressed air within. What is true of air is true of all gases. Dangerous boiler explosions have occurred because the boiler walls were not strong enough to withstand the pressure of the steam (which is water in the form of gas). The pressure within the boilers of engines is frequently several hundred pounds to the square inch, and such a pressure needs a strong boiler.

87. How Pressure is Measured in Buildings. In the preceding Section we saw that undue pressure of a gas may cause explosion. It is important, therefore, that authorities keep strict watch on gases confined within pipes and reservoirs, never allowing the pressure to exceed that which the walls of the reservoir will safely bear.

[Illustration: FIG. 52.--A pressure gauge.]

CHAPTER VIII

Pressure in a gas pipe may be measured by a simple instrument called the pressure gauge: The gauge consists of a bent glass tube containing mercury, and so made that one end can be fitted to a gas jet (Fig. 52). When the gas cock is closed, the mercury stands at the same level in both arms, but when the cock is opened, the gas whose pressure is being measured forces the mercury up the opposite arm. If the pressure of the gas is small, the mercury changes its level but very little. It is clear that the height of a column of mercury is a measure of the gas pressure. Now it is known that one cubic inch of mercury weighs about half a pound. Hence a column of mercury one inch high indicates a pressure of about one half pound to the square inch; a column two inches high indicates a pressure of about one pound to the square inch, and so on.

This is a very convenient way to measure the pressure of the illuminating gas in our homes and offices. The gauge is attached to the gas burner and the pressure is read by means of a scale attached to the gauge. (See Laboratory Manual.)

In order to have satisfactory illumination, the pressure must be strong enough to give a steady, broad flame. If the flame from any gas jet is flickering and weak, it is usually an indication of insufficient pressure and the gas company should investigate conditions and see to it that the consumer receives his proper value.

87. The Gas Meter. Most householders are deeply interested in the actual amount of gas which they consume (gas is charged for according to the number of cubic feet used), and therefore they should be able to read the gas meter which indicates their consumption of gas. Such gas meters are furnished by the companies, and can be read easily.

[Illustration: FIG. 53.--The gas meter indicates the number of cubic feet of gas consumed.]

The instrument itself is somewhat complex. It will suffice to say that within the meter box are thin disks which are moved by the stream of gas that passes them. This movement of the disks is recorded by clockwork devices on a dial face. In this way, the number of cubic feet

CHAPTER VIII 69

of gas which pass through the meter is automatically registered.

89. **The Relation between Pressure and Volume.** It was long known that as the pressure of a gas increases, that is, as it becomes compressed, its volume decreases, but Robert Boyle was the first to determine the exact relation between the volume and the pressure of a gas. He did this in a very simple manner.

Pour mercury into a U-shaped tube until the level of the mercury in the closed end of the tube is the same as the level in the open end. The air in the long arm is pressing upon the mercury in that arm, and is tending to force it up the short arm. The air in the short closed arm is pressing down upon the mercury in that arm and tending to send it up the long arm. Since the mercury is at the same level in the two arms, the pressure in the long arm must be equal to the pressure in the short arm. But the long arm is open, and the pressure in that arm is the pressure of the atmosphere. Therefore the pressure in the short arm must be one atmosphere. Measure the distance *bc* between the top of the mercury and the closed end of the tube.

[Illustration: FIGS. 54, 55.--As the pressure on the gas increases, its volume decreases.]

Pour more mercury into the open end of the tube, and as the mercury rises higher and higher in the long arm, note carefully the decrease in the volume of the air in the short arm. Pour mercury into the tube until the difference in level *bd* is just equal to the barometric height, approximately 32 inches. The pressure of the air in the closed end now supports the pressure of one atmosphere, and in addition, a column of mercury equal to another atmosphere. If now the air column in the closed end is measured, its volume will be only one half of its former volume. By doubling the pressure we have reduced the volume one half. Similarly, if the pressure is increased threefold, the volume will be reduced to one third of the original volume.

90. **Heat due to Compression.** We saw in Section 89 that whenever the pressure exerted upon a gas is increased, the volume of the gas is decreased; and that whenever the pressure upon a gas is decreased, the volume of the gas is increased. If the pressure is changed very

CHAPTER VIII

slowly, the change in the temperature of the gas is imperceptible; if, however, the pressure is removed suddenly, the temperature falls rapidly, or if the pressure is applied suddenly, the temperature rises rapidly. When bicycle tires are being inflated, the pump becomes hot because of the compression of the air.

The amount of heat resulting from compression is surprisingly large; for example, if a mass of gas at 0° C. is suddenly compressed to one half its original volume, its temperature rises 87° C.

91. **Cooling by Expansion.** If a gas expands suddenly, its temperature falls; for example, if a mass of gas at 87° C. is allowed to expand rapidly to twice its original volume, its temperature falls to 0° C. If the compressed air of a bicycle tire is allowed to expand and a sensitive thermometer is held in the path of the escaping air, the thermometer will show a decided drop in temperature.

The low temperature obtained by the expansion of air or other gases is utilized commercially on a large scale. By means of powerful pistons air is compressed to one third or one fourth its original volume, is passed through a coil of pipe surrounded with cold water, and is then allowed to escape into large refrigerating vaults, which thereby have their temperatures noticeably lowered, and can be used for the permanent storage of meats, fruits, and other perishable material. In summer, when the atmospheric temperature is high, the storage and preservation of foods is of vital importance to factories and cold storage houses, and but for the low temperature obtainable by the expansion of compressed gases, much of our food supply would be lost to use.

92. **Unexpected Transformations.** If the pressure on a gas is greatly increased, a sudden transformation sometimes occurs and the gas becomes a liquid. Then, if the pressure is reduced, a second transformation occurs, and the liquid evaporates or returns to its original form as a gas.

In Section 23 we saw that a fall of temperature caused water vapor to condense or liquefy. If temperature alone were considered, most gases could not be liquefied, because the temperature at which the average gas liquefies is so low as to be out of the range of possibility; it has been calculated, for example, that a temperature of 252° C. below zero would have to be obtained in order to liquefy hydrogen.

Some gases can be easily transformed into liquids by pressure alone, some gases can be easily transformed into liquids by cooling alone; on the other hand, many gases are so difficult to liquefy that both pressure and low temperature are needed to produce the desired result. If a gas is cooled and compressed at the same time, liquefaction occurs much more surely and easily than though either factor alone were depended upon. The air which surrounds us, and of whose existence we are scarcely aware, can be reduced to the form of a liquid, but the pressure exerted upon the portion to be liquefied must be thirty-nine times as great as the atmospheric pressure, and the temperature must have been reduced to a very low point.

93. Artificial Ice. Ammonia gas is liquefied by strong pressure and low temperature and is then allowed to flow into pipes which run through tanks containing salt water. The reduction of pressure causes the liquid to evaporate or turn to a gas, and the fall of temperature which always accompanies evaporation means a lowering of the temperature of the salt water to 16° or 18° below zero. But immersed in the salt water are molds containing pure water, and since the freezing point of water is 0° C, the water in the molds freezes and can be drawn from the mold as solid cakes of ice.

[Illustration: FIG. 56.--Apparatus for making artificial ice.]

Ammonia gas is driven by the pump *C* into the coil *D* (Fig. 56) under a pressure strong enough to liquefy it, the heat generated by this compression being carried off by cold water which constantly circulates through *B*. The liquid ammonia flows through the regulating valve *V* into the coil *E*, in which the pressure is kept low by the pump *C*. The accompanying expansion reduces the temperature to a very low degree, and the brine which circulates around the coil *E* acquires a temperature below the freezing point of pure water. The cold brine passes from *A* to a tank in which are immersed cans filled with water, and within a short time the water in the cans is frozen into solid cakes of ice.

CHAPTER IX

INVISIBLE OBJECTS

94. Very Small Objects. We saw in Section 84 that gases have a tendency to expand, but that they can be compressed by the application of force. This observation has led scientists to suppose that substances are composed of very minute particles called molecules, separated by small spaces called pores; and that when a gas is condensed, the pores become smaller, and that when a gas expands, the pores become larger.

The fact that certain substances are soluble, like sugar in water, shows that the molecules of sugar find a lodging place in the spaces or pores between the molecules of water, in much the same way that pebbles find lodgment in the chinks of the coal in a coal scuttle. An indefinite quantity of sugar cannot be dissolved in a given quantity of liquid, because after a certain amount of sugar has been dissolved all the pores become filled, and there is no available molecular space. The remainder of the sugar settles at the bottom of the vessel, and cannot be dissolved by any amount of stirring.

If a piece of potassium permanganate about the size of a grain of sand is put into a quart of water, the solid disappears and the water becomes a deep rich red. The solid evidently has dissolved and has broken up into minute particles which are too small to be seen, but which have scattered themselves and lodged in the pores of the water, thus giving the water its rich color.

There is no visible proof of the existence of molecules and molecular spaces, because not only are our eyes unable to see them directly, but even the most powerful microscope cannot make them visible to us. They are so small that if one thousand of them were laid side by side, they would make a speck too small to be seen by the eye and too small to be visible under the most powerful microscope.

We cannot see molecules or molecular pores, but the phenomena of compression and expansion, solubility and other equally convincing facts, have led us to conclude that all substances are composed of

CHAPTER IX   73

very minute particles or molecules separated by spaces called pores.

95. Journeys Made by Molecules. If a gas jet is turned on and not lighted, an odor of gas soon becomes perceptible, not only throughout the room, but in adjacent halls and even in distant rooms. An uncorked bottle of cologne scents an entire room, the odor of a rose or violet permeates the atmosphere near and far. These simple everyday occurrences seem to show that the molecules of a gas must be in a state of continual and rapid motion. In the case of the cologne, some molecules must have escaped from the liquid by the process of evaporation and traveled through the air to the nose. We know that the molecules of a liquid are in motion and are continually passing into the air because in time the vessel becomes empty. The only way in which this could happen would be for the molecules of the liquid to pass from the liquid into the surrounding medium; but this is really saying that the molecules are in motion.

From these phenomena and others it is reasonably clear that substances are composed of molecules, and that molecules are not inert, quiet particles, but that they are in incessant motion, moving rapidly hither and thither, sometimes traveling far, sometimes near. Even the log of wood which lies heavy and motionless on our woodpile is made up of countless billions of molecules each in rapid incessant motion. The molecules of solid bodies cannot escape so readily as those of liquids and gases, and do not travel far. The log lies year after year in an apparently motionless condition, but if one's eyes were keen enough, the molecules would be seen moving among themselves, even though they cannot escape into the surrounding medium and make long journeys as do the molecules of liquids and gases.

96. The Companions of Molecules. Common sense tells us that a molecule of water is not the same as a molecule of vinegar; the molecules of each are extremely small and in rapid motion, but they differ essentially, otherwise one substance would be like every other substance. What is it that makes a molecule of water differ from a molecule of vinegar, and each differ from all other molecules? Strange to say, a molecule is not a simple object, but is quite complex, being composed of one or more smaller particles, called atoms, and the number and kind of atoms in a molecule determine the type of the

CHAPTER IX 74

molecule, and the type of the molecule determines the substance. For example, a glass of water is composed of untold millions of molecules, and each molecule is a company of three still smaller particles, one of which is called the oxygen atom and two of which are alike in every particular and are called hydrogen atoms.

97. Simple Molecules. Generally molecules are composed of atoms which are different in kind. For example, the molecule of water has two different atoms, the oxygen atom and the hydrogen atoms; alcohol has three different kinds of atoms, oxygen, hydrogen, and carbon. Sometimes, however, molecules are composed of a group of atoms all of which are alike. Now there are but seventy or eighty different kinds of atoms, and hence there can be but seventy or eighty different substances whose molecules are composed of atoms which are alike. When the atoms comprising a molecule are all alike, the substance is called an element, and is said to be a simple substance. Throughout the length and breadth of this vast world of ours there are only about eighty known elements. An element is the simplest substance conceivable, because it has not been separated into anything simpler. Water is a compound substance. It can be separated into oxygen and hydrogen.

Gold, silver, and lead are examples of elements, and water, alcohol, cider, sand, and marble are complex substances, or compounds, as we are apt to call them. Everything, no matter what its size or shape or character, is formed from the various combinations into molecules of a few simple atoms, of which there exist about eighty known different kinds. But few of the eighty known elements play an important part in our everyday life. The elements in which we are most interested are given in the following table, and the symbols by which they are known are placed in columns to the right:

Oxygen	O	Copper	Cu	Phosphorus	P	Hydrogen	H	Iodine	I
Potassium	K	Carbon	C	Iron	Fe	Silver	Ag	Aluminium	Al	Lead	Pb
Sodium	Na			Calcium	Ca	Nickel	Ni	Sulphur	S	Chlorine	Cl
Nitrogen	N	Tin	Sn								

We have seen in an earlier experiment that twice as much hydrogen as oxygen can be obtained from water. Two atoms of the element hydrogen unite with one atom of the element oxygen to make one molecule of water. In symbols we express this H_2O. A group of symbols, such as this, expressing a molecule of a compound is called a *formula*. NaCl is the formula for sodium chloride, which is the chemical name of common salt.

CHAPTER X

LIGHT

98. What Light Does for Us. Heat keeps us warm, cooks our food, drives our engines, and in a thousand ways makes life comfortable and pleasant, but what should we do without light? How many of us could be happy even though warm and well fed if we were forced to live in the dark where the sunbeams never flickered, where the shadows never stole across the floor, and where the soft twilight could not tell us that the day was done? Heat and light are the two most important physical factors in life; we cannot say which is the more necessary, because in the extreme cold or arctic regions man cannot live, and in the dark places where the light never penetrates man sickens and dies. Both heat and light are essential to life, and each has its own part to play in the varied existence of man and plant and animal.

Light enables us to see the world around us, makes the beautiful colors of the trees and flowers, enables us to read, is essential to the taking of photographs, gives us our moving pictures and our magic lanterns, produces the exquisite tints of stained-glass windows, and brings us the joy of the rainbow. We do not always realize that light is beneficial, because sometimes it fades our clothing and our carpets, and burns our skin and makes it sore. But we shall see that even these apparently harmful effects of light are in reality of great value in man's constant battle against disease.

99. The Candle. Natural heat and light are furnished by the sun, but the absence of the sun during the evening makes artificial light necessary, and even during the day artificial light is needed in

CHAPTER X

buildings whose structure excludes the natural light of the sun. Artificial light is furnished by electricity, by gas, by oil in lamps, and in numerous other ways. Until modern times candles were the main source of light, and indeed to-day the intensity, or power, of any light is measured in candle power units, just as length is measured in yards; for example, an average gas jet gives a 10 candle power light, or is ten times as bright as a candle; an ordinary incandescent electric light gives a 16 candle power light, or furnishes sixteen times as much light as a candle. Very strong large oil lamps can at times yield a light of 60 candle power, while the large arc lamps which flash out on the street corners are said to furnish 1200 times as much light as a single candle. Naturally all candles do not give the same amount of light, nor are all candles alike in size. The candles which decorate our tea tables are of wax, while those which serve for general use are of paraffin and tallow.

[Illustration: FIG. 57.--A photograph at *a* receives four times as much light as when held at *b*.]

100. Fading Illumination. The farther we move from a light, the less strong, or intense, is the illumination which reaches us; the light of the street lamp on the corner fades and becomes dim before the middle of the block is reached, so that we look eagerly for the next lamp. The light diminishes in brightness much more rapidly than we realize, as the following simple experiment will show. Let a single candle (Fig. 57) serve as our light, and at a distance of one foot from the candle place a photograph. In this position the photograph receives a definite amount of light from the candle and has a certain brightness.

If now we place a similar photograph directly behind the first photograph and at a distance of two feet from the candle, the second photograph receives no light because the first one cuts off all the light. If, however, the first photograph is removed, the light which fell on it passes outward and spreads itself over a larger area, until at the distance of the second photograph the light spreads itself over four times as large an area as formerly. At this distance, then, the illumination on the second photograph is only one fourth as strong as it was on a similar photograph held at a distance of one foot from the candle.

CHAPTER X   77

The photograph or object placed at a distance of one foot from a light is well illuminated; if it is placed at a distance of two feet, the illumination is only one fourth as strong, and if the object is placed three feet away, the illumination is only one ninth as strong. This fact should make us have thought and care in the use of our eyes. We think we are sixteen times as well off with our incandescent lights as our ancestors were with simple candles, but we must reflect that our ancestors kept the candle near them, "at their elbow," so to speak, while we sit at some distance from the light and unconcernedly read and sew.

As an object recedes from a light the illumination which it receives diminishes rapidly, for the strength of the illumination is inversely proportional to the square of distance of the object from the light. Our ancestors with a candle at a distance of one foot from a book were as well off as we are with an incandescent light four feet away.

101. Money Value of Light. Light is bought and sold almost as readily as are the products of farm and dairy; many factories, churches, and apartments pay a definite sum for electric light of a standard strength, and naturally full value is desired. An instrument for measuring the strength of a light is called a photometer, and there are many different varieties, just as there are varieties of scales which measure household articles. One light-measuring scale depends upon the law that the intensity of illumination decreases with the square of the distance of the object from the light. Suppose we wish to measure the strength of the electric light bulbs in our homes, in order to see whether we are getting the specified illumination. In front of a screen place a black rod (Fig. 58) which is illuminated by two different lights; namely, a standard candle and an incandescent bulb whose strength is to be measured. Two shadows of the rod will fall on the screen, one caused by the candle and the other caused by the incandescent light. The shadow due to the latter source is not so dark as that due to the candle. Now let the incandescent light be moved away from the screen until the two shadows are of equal darkness. If the incandescent light is four times as far away from the screen as the candle, and the shadows are equal, we know, by Section 100, that its strength is sixteen candle power. If the incandescent light is four times as far away from the screen as the candle is, its power must be sixteen times

CHAPTER X   78

as great, and we know the company is furnishing the standard amount of light for a sixteen candle power electric bulb. If, however, the bulb must be moved nearer to the rod in order that the two shadows may be similar then the light given by the bulb is less than sixteen candle power, and less than that due the consumer.

[Illustration: FIG. 58.--The two shadows are equally dark.]

102. How Light Travels. We never expect to see around a corner, and if we wish to see through pinholes in three separate pieces of cardboard, we place the cardboards so that the three holes are in a straight line. When sunlight enters a dark room through a small opening, the dust particles dancing in the sun show a straight ray. If a hole is made in a card, and the card is held in front of a light, the card casts a shadow, in the center of which is a bright spot. The light, the hole, and the bright spot are all in the same straight line. These simple observations lead us to think that light travels in a straight line.

[Illustration: FIG. 59.--The candle cannot be seen unless the three pinholes are in a strait line.]

We can always tell the direction from which light comes, either by the shadow cast or by the bright spot formed when an opening occurs in the opaque object casting the shadow. If the shadow of a tree falls towards the west, we know the sun must be in the cast; if a bright spot is on the floor, we can easily locate the light whose rays stream through an opening and form the bright spot. We know that light travels in a straight line, and following the path of the beam which comes to our eyes, we are sure to locate the light.

103. Good and Bad Mirrors. As we walk along the street, we frequently see ourselves reflected in the shop windows, in polished metal signboards, in the metal trimmings of wagons and automobiles; but in mirrors we get the best image of ourselves. We resent the image given by a piece of tin, because the reflection is distorted and does not picture us as we really are; a rough surface does not give a fair representation; if we want a true image of ourselves, we must use a smooth surface like a mirror as a reflector. If the water in a pond is absolutely still, we get a clear, true image of the trees, but if there are

CHAPTER X

ripples on the surface, the reflection is blurred and distorted. A metal roof reflects so much light that the eyes are dazzled by it, and a whitewashed fence injures the eyes because of the glare which comes from the reflected light. Neither of these could be called mirrors, however, because although they reflect light, they reflect it so irregularly that not even a suggestion of an image can be obtained.

Most of us are sufficiently familiar with mirrors to know that the image is a duplicate of ourselves with regard to size, shape, color, and expression, but that it appears to be back of the mirror, while we are actually in front of the mirror. The image appears not only behind the mirror, but it is also exactly as far back of the mirror as we are in front of it; if we approach the mirror, the image also draws nearer; if we withdraw, it likewise recedes.

104. The Path of Light. If a mirror or any other polished surface is held in the path of a sunbeam, some of the light is reflected, and by rotating the mirror the reflected sunbeam may be made to take any path. School children amuse themselves by reflecting sunbeams from a mirror into their companions' faces. If the companion moves his head in order to avoid the reflected beam, his tormentor moves or inclines the mirror and flashes the beam back to his victim's face.

If a mirror is held so that a ray of light strikes it in a perpendicular direction, the light is reflected backward along the path by which it came. If, however, the light makes an angle with the mirror, its direction is changed, and it leaves the mirror along a new path. By observation we learn that when a beam strikes the mirror and makes an angle of 30° with the perpendicular, the beam is reflected in such a way that its new path also makes an angle of 30° with the perpendicular. If the sunbeam strikes the mirror at an angle of 32° with the perpendicular, the path of the reflected ray also makes an angle of 32° with the perpendicular. The ray (*AC*, Fig. 60) which falls upon the mirror is called the incident ray, and the angle which the incident ray (*AC*) makes with the perpendicular (*BC*) to the mirror, at the point where the ray strikes the mirror, is called the angle of incidence. The angle formed by the reflected ray (*CD*) and this same perpendicular is called the angle of reflection. Observation and experiment have taught us that light is always reflected in such a way that the angle of

CHAPTER X 80

reflection equals the angle of incidence. Light is not the only illustration we have of the law of reflection. Every child who bounces a ball makes use of this law, but he uses it unconsciously. If an elastic ball is thrown perpendicularly against the floor, it returns to the sender; if it is thrown against the floor at an angle (Fig. 61), it rebounds in the opposite direction, but always in such a way that the angle of reflection equals the angle of incidence.

[Illustration: FIG. 60.--The ray *AC* is reflected as *CD*.]

[Illustration: FIG. 61.--A bouncing ball illustrates the law of reflection.]

105. Why the Image seems to be behind the Mirror. If a candle is placed in front of a mirror, as in Figure 62, one of the rays of light which leaves the candle will fall upon the mirror as *AB* and will be reflected as *BC* (in such a way that the angle of reflection equals the angle of incidence). If an observer stands at *C*, he will think that the point *A* of the candle is somewhere along the line *CB* extended. Such a supposition would be justified from Section 102. But the candle sends out light in all directions; one ray therefore will strike the mirror as *AD* and will be reflected as *DE*, and an observer at *E* will think that the point *A* of the candle is somewhere along the line *ED*. In order that both observers may be correct, that is, in order that the light may seem to be in both these directions, the image of the point *A* must seem to be at the intersection of the two lines. In a similar manner it can be shown that every point of the image of the candle seems to be behind the mirror.

[Illustration: FIG. 62.--The image is a duplicate of the object, but appears to be behind the mirror.]

It can be shown by experiment that the distance of the image behind the mirror is equal to the distance of the object in front of the mirror.

106. Why Objects are Visible. If the beam of light falls upon a sheet of paper, or upon a photograph, instead of upon a smooth polished surface, no definite reflected ray will be seen, but a glare will be produced by the scattering of the beam of light. The surface of the paper or photograph is rough, and as a result, it scatters the beam in

every direction. It is hard for us to realize that a smooth sheet of paper is by no means so smooth as it looks. It is rough compared with a polished mirror. The law of reflection always holds, however, no matter what the reflecting surface is,--the angle of reflection always equals the angle of incidence. In a smooth body the reflected beams are all parallel; in a rough body, the reflected beams are inclined to each other in all sorts of ways, and no two beams leave the paper in exactly the same direction.

[Illustration: FIG. 63.--The surface of the paper, although smooth in appearance, is in reality rough, and scatters the light in every direction.]

Hot coals, red-hot stoves, gas flames, and candles shine by their own light, and are self-luminous. Objects like chairs, tables, carpets, have no light within themselves and are visible only when they receive light from a luminous source and reflect that light. We know that these objects are not self-luminous, because they are not visible at night unless a lamp or gas is burning. When light from any luminous object falls upon books, desks, or dishes, it meets rough surfaces, and hence undergoes diffuse reflection, and is scattered irregularly in all directions. No matter where the eye is, some reflected rays enter it, and the various objects are clearly seen.

CHAPTER XI

REFRACTION

107. Bent Rays of Light. A straw in a glass of lemonade seems to be broken at the surface of the liquid, the handle of a teaspoon in a cup of water appears broken, and objects seen through a glass of water may seem distorted and changed in size. When light passes from air into water, or from any transparent substance into another of different density, its direction is changed, and it emerges along an entirely new path (Fig. 64). We know that light rays pass through glass, because we can see through the window panes and through our spectacles; we know that light rays pass through water, because we can see through

CHAPTER XI

a glass of clear water; on the other hand, light rays cannot pass through wood, leather, metal, etc.

[Illustration: FIG. 64.--A straw or stick in water seems broken.]

Whenever light meets a transparent substance obliquely, some of it is reflected, undergoing a change in its direction; and some of it passes onward through the medium, but the latter portion passes onward along a new path. The ray *RO* (Fig. 65) passes obliquely through the air to the surface of the water, but, on entering the water, it is bent or refracted and takes the new path *OS*. The angle *AOR* is called the angle of incidence. The angle *POS* is called the angle of refraction.

[Illustration: FIG. 65.--When the ray *RO* enters the water, its path changes to *OS*.]

The angle of refraction is the angle formed by the refracted ray and the perpendicular to the surface at the point where the light strikes it.

When light passes from air into water or glass, the refracted ray is bent toward the perpendicular, so that the angle of refraction is smaller than the angle of incidence. When a ray of light passes from water or glass into air, the refracted ray is bent away from the perpendicular so that the angle of refraction is greater than the angle of incidence.

The bending or deviation of light in its passage from one substance to another is called refraction.

108. How Refraction Deceives us. Refraction is the source of many illusions; bent rays of light make objects appear where they really are not. A fish at *A* (Fig. 66) seems to be at *B*. The end of the stick in Figure 64 seems to be nearer the surface of the water than it really is.

[Illustration: FIG. 66.--A fish at *A* seems to be at *B*.]

The light from the sun, moon, and stars can reach us only by passing through the atmosphere, but in Section 76, we learned that the atmosphere varies in density from level to level; hence all the light which travels through the atmosphere is constantly deviated from its

CHAPTER XI 83

original path, and before the light reaches the eye it has undergone many changes in direction. Now we learned in Section 102, that the direction of the rays of light as they enter the eye determines the direction in which an object is seen; hence the sun, moon, and stars seem to be along the lines which enter the eye, although in reality they are not.

109. Uses of Refraction. If it were not for refraction, or the deviation of light in its passage from medium to medium, the wonders and beauties of the magic lantern and the camera would be unknown to us; sun, moon, and stars could not be made to yield up their distant secrets to us in photographs; the comfort and help of spectacles would be lacking, spectacles which have helped unfold to many the rare beauties of nature, such as a clear view of clouds and sunset, of humming bee and flying bird. Books with their wealth of entertainment and information would be sealed to a large part of mankind, if glasses did not assist weak eyes.

By refraction the magnifying glass reveals objects hidden because of their minuteness, and enlarges for our careful contemplation objects otherwise barely visible. The watchmaker, unassisted by the magnifying glass, could not detect the tiny grains of dust or sand which clog the delicate wheels of our watches. The merchant, with his lens, examines the separate threads of woolen and silk fabrics to determine the strength and value of the material. The physician, with his invaluable microscope, counts the number of infinitesimal corpuscles in the blood and bases his prescription on that count; he examines the sputum of a patient to determine whether tuberculosis wastes the system. The bacteriologist with the same instrument scrutinizes the drinking water and learns whether the dangerous typhoid germs are present. The future of medicine will depend somewhat upon the additional secrets which man is able to force from nature through the use of powerful lenses, because as lenses have, in the past, been the means of revealing disease germs, so in the future more powerful lenses may serve to bring to light germs yet unknown. How refraction accomplishes these results will be explained in the following Sections.

110. The Window Pane. We have seen that light is bent when it passes from one medium to another of different density, and that

CHAPTER XI

objects viewed by refracted light do not appear in their proper positions.

When a ray of light passes through a piece of plane glass, such as a window pane (Fig. 67), it is refracted at the point B toward the perpendicular, and continues its course through the glass in the new direction BC. On emerging from the glass, the light is refracted away from the perpendicular and takes the direction CD, which is clearly parallel to its original direction. Hence, when we view objects through the window, we see them slightly displaced in position, but otherwise unchanged. The deviation or displacement caused by glass as thin as window panes is too slight to be noticed, and we are not conscious that objects are out of position.

[Illustration: FIG. 67.--Objects looked at through a window pane seem to be in their natural place.]

111. Chandelier Crystals and Prisms. When a ray of light passes through plane glass, like a window pane, it is shifted somewhat, but its direction does not change; that is, the emergent ray is parallel to the incident ray. But when a beam of light passes through a triangular glass prism, such as a chandelier crystal, its direction is greatly changed, and an object viewed through a prism is seen quite out of its true position.

Whenever light passes through a prism, it is bent toward the base of the prism, or toward the thick portion of the prism, and emerges from the prism in quite a different direction from that in which it entered (Fig. 68). Hence, when an object is looked at through a prism, it is seen quite out of place. In Figure 68, the candle seems to be at S, while in reality it is at A.

[Illustration: FIG. 68.--When looked at through the prism, A seems to be at S.]

112. Lenses. If two prisms are arranged as in Figure 69, and two parallel rays of light fall upon the prisms, the beam A will be bent downward toward the thickened portion of the prism, and the beam B will be bent upward toward the thick portion of the prism, and after

CHAPTER XI

passing through the prism the two rays will intersect at some point *F*, called a focus.

[Illustration: FIG. 69.--Rays of light are converged and focused at *F*.]

If two prisms are arranged as in Figure 70, the ray *A* will be refracted upward toward the thick end, and the ray *B* will be refracted downward toward the thick end; the two rays, on emerging, will therefore be widely separated and will not intersect.

[Illustration: FIG. 70.--Rays of light are diverged and do not come to any real focus.]

Lenses are very similar to prisms; indeed, two prisms placed as in Figure 69, and rounded off, would make a very good convex lens. A lens is any transparent material, but usually glass, with one or both sides curved. The various types of lenses are shown in Figure 71.

[Illustration: FIG. 71.--The different types of lenses.]

The first three types focus parallel rays at some common point *F*, as in Figure 69. Such lenses are called convex or converging lenses. The last three types, called concave lenses, scatter parallel rays so that they do not come to a focus, but diverge widely after passage through the lens.

113. **The Shape and Material of a Lens.** The main or principal focus of a lens, that is, the point at which rays parallel to the base line *AB* meet (Fig. 71), depends upon the shape of the lens. For example, a thick lens, such as *A* (Fig. 72), focuses the rays very near to the lens; *B*, which is not so thick, focuses the rays at a greater distance from the lens; and *C*, which is a very thin lens, focuses the rays at a considerable distance from the lens. The distance of the principal focus from the lens is called the focal length of the lens, and from the diagrams we see that the more convex the lens, the shorter the focal length.

[Illustration: FIG. 72.--The more curved the lens, the shorter the focal length, and the nearer the focus is to the lens.]

CHAPTER XI

The position of the principal focus depends not only on the shape of the lens, but also on the refractive power of the material composing the lens. A lens made of ice would not deviate the rays of light so much as a lens of similar shape composed of glass. The greater the refractive power of the lens, the greater the bending, and the nearer the principal focus to the lens.

There are many different kinds of glass, and each kind of glass refracts the light differently. Flint glass contains lead; the lead makes the glass dense, and gives it great refractive power, enabling it to bend and separate light in all directions. Cut glass and toilet articles are made of flint glass because of the brilliant effects caused by its great refractive power, and imitation gems are commonly nothing more than polished flint glass.

114. How Lenses Form Images. Suppose we place an arrow, *A*, in front of a convex lens (Fig. 73). The ray *AC*, parallel to the principal axis, will pass through the lens and emerge as *DE*. The ray is always bent toward the thick portion of the lens, both at its entrance into the lens and its emergence from the lens.

[Illustration: FIG. 73.--The image is larger than the object. By means of a lens, a watchmaker gets an enlarged image of the dust which clogs the wheels of his watch.]

In Section 105, we saw that two rays determine the position of any point of our image; hence in order to locate the image of the top of the arrow, we need to consider but one more ray from the top of the object. The most convenient ray to choose would be one passing through *O*, the optical center of the lens, because such a ray passes through the lens unchanged in direction, as is clear from Figure 74. The point where *AC* and *AO* meet after refraction will be the position of the top of the arrow. Similarly it can be shown that the center of the arrow will be at the point *T*, and we see that the image is larger than the object. This can be easily proved experimentally. Let a convex lens be placed near a candle (Fig. 75); move a paper screen back and forth behind the lens; for some position of the screen a clear, enlarged image of the candle will be made.

CHAPTER XI   87

[Illustration: FIG. 74.--Rays above *O* are bent downward, those below *O* are bent upward, and rays through *O* emerge from the lens unchanged in direction.]

If the candle or arrow is placed in a new position, say at *MA* (Fig. 76), the image formed is smaller than the object, and is nearer to the lens than it was before. Move the lens so that its distance from the candle is increased, and then find the image on a piece of paper. The size and position of the image depend upon the distance of the object from the lens (Fig. *77*). By means of a lens one can easily get on a visiting card a picture of a distant church steeple.

[Illustration: FIG. 75.--The lens is held in such a position that the image of the candle is larger than the object.]

[Illustration: FIG. 76.--The image is smaller than the object.]

115. The Value of Lenses. If it were not for the fact that a lens can be held at such a distance from an object as to make the image larger than the object, it would be impossible for the lens to assist the watchmaker in locating the small particles of dust which clog the wheels of the watch. If it were not for the opposite fact--that a lens can be held at such a distance from the object as to make an image smaller than the object, it would be impossible to have a photograph of a tall tree or building unless the photograph were as large as the tree itself. When a photographer takes a photograph of a person or a tree, he moves his camera until the image formed by the lens is of the desired size. By bringing the camera (really the lens of the camera) near, we obtain a large-sized photograph; by increasing the distance between the camera and the object, a smaller photograph is obtained. The mountain top may be so far distant that in the photograph it will not appear to be greater than a small stone.

[Illustration: FIG. 77.--The lens is placed in such a position that the image is about the same size as the object.]

Many familiar illustrations of lenses, or curved refracting surfaces, and their work, are known to all of us. Fish globes magnify the fish that swim within. Bottles can be so shaped that they make the olives,

CHAPTER XI 88

pickles, and peaches that they contain appear larger than they really are. The fruit in bottles frequently seems too large to have gone through the neck of the bottle. The deception is due to refraction, and the material and shape of the bottle furnish a sufficient explanation.

By using combinations of two or more lenses of various kinds, it is possible to have an image of almost any desired size, and in practically any desired position.

116. The Human Eye. In Section 114, we obtained on a movable screen, by means of a simple lens, an image of a candle. The human eye possesses a most wonderful lens and screen (Fig. 78); the lens is called the crystalline lens, and the screen is called the retina. Rays of light pass from the object through the pupil P, go through the crystalline lens L, where they are refracted, and then pass onward to the retina R, where they form a distinct image of the object.

[Illustration: FIG. 78.--The eye.]

We learned in Section 114 that a change in the position of the object necessitated a change in the position of the screen, and that every time the object was moved the position of the screen had to be altered before a clear image of the object could be obtained. The retina of the eye cannot be moved backward and forward, as the screen was, and the crystalline lens is permanently located directly back of the iris. How, then, does it happen that we can see clearly both near and distant objects; that the printed page which is held in the hand is visible at one second, and that the church spire on the distant horizon is visible the instant the eyes are raised from the book? How is it possible to obtain on an immovable screen by means of a simple lens two distinct images of objects at widely varying distances?

The answer to these questions is that the crystalline lens changes shape according to need. The lens is attached to the eye by means of small muscles, m, and it is by the action of these muscles that the lens is able to become small and thick, or large and thin; that is, to become more or less curved. When we look at near objects, the muscles act in such a way that the lens bulges out, and becomes thick in the middle and of the right curvature to focus the near object upon the screen.

CHAPTER XI   89

When we look at an object several hundred feet away, the muscles change their pull on the lens and flatten it until it is of the proper curvature for the new distance. The adjustment of the muscles is so quick and unconscious that we normally do not experience any difficulty in changing our range of view. The ability of the eye to adjust itself to varying distances is called accommodation. The power of adjustment in general decreases with age.

117. Farsightedness and Nearsightedness. A farsighted person is one who cannot see near objects so distinctly as far objects, and who in many cases cannot see near objects at all. The eyeball of a farsighted person is very short, and the retina is too close to the crystalline lens. Near objects are brought to a focus behind the retina instead of on it, and hence are not visible. Even though the muscles of accommodation do their best to bulge and thicken the lens, the rays of light are not bent sufficiently to focus sharply on the retina. In consequence objects look blurred. Farsightedness can be remedied by convex glasses, since they bend the light and bring it to a closer focus. Convex glasses, by bending the rays and bringing them to a nearer focus, overbalance a short eyeball with its tendency to focus objects behind the retina.

[Illustration: FIG. 79.--The farsighted eye.]

[Illustration: FIG. 80.--The defect is remedied by convex glasses.]

A nearsighted person is one who cannot see objects unless they are close to the eye. The eyeball of a nearsighted person is very wide, and the retina is too far away from the crystalline lens. Far objects are brought to a focus in front of the retina instead of on it, and hence are not visible. Even though the muscles of accommodation do their best to pull out and flatten the lens, the rays are not separated sufficiently to focus as far back as the retina. In consequence objects look blurred. Nearsightedness can be remedied by wearing concave glasses, since they separate the light and move the focus farther away. Concave glasses, by separating the rays and making the focus more distant, overbalance a wide eyeball with its tendency to focus objects in front of the retina.

[Illustration: FIG. 81.--The nearsighted eye. The defect is remedied by concave glasses.]

118. Headache and Eyes. Ordinarily the muscles of accommodation adjust themselves easily and quickly; if, however, they do not, frequent and severe headaches occur as a result of too great muscular effort toward accommodation. Among young people headaches are frequently caused by over-exertion of the crystalline muscles. Glasses relieve the muscles of the extra adjustment, and hence are effective in eliminating this cause of headache.

An exact balance is required between glasses, crystalline lens, and muscular activity, and only those who have studied the subject carefully are competent to treat so sensitive and necessary a part of the body as the eye. The least mistake in the curvature of the glasses, the least flaw in the type of glass (for example, the kind of glass used), means an improper focus, increased duty for the muscles, and gradual weakening of the entire eye, followed by headache and general physical discomfort.

119. Eye Strain. The extra work which is thrown upon the nervous system through seeing, reading, writing, and sewing with defective eyes is recognized by all physicians as an important cause of disease. The tax made upon the nervous system by the defective eye lessens the supply of energy available for other bodily use, and the general health suffers. The health is improved when proper glasses are prescribed.

Possibly the greatest danger of eye strain is among school children, who are not experienced enough to recognize defects in sight. For this reason, many schools employ a physician who examines the pupils' eyes at regular intervals.

The following general precautions are worth observing:--

1. Rest the eyes when they hurt, and as far as possible do close work, such as writing, reading, sewing, wood carving, etc., by daylight.

2. Never read in a very bright or a very dim light.

3. If the light is near, have it shaded.

4. Do not rub the eyes with the fingers.

5. If eyes are weak, bathe them in lukewarm water in which a pinch of borax has been dissolved.

CHAPTER XII

PHOTOGRAPHY

120. The Magic of the Sun. Ribbons and dresses washed and hung in the sun fade; when washed and hung in the shade, they are not so apt to lose their color. Clothes are laid away in drawers and hung in closets not only for protection against dust, but also against the well-known power of light to weaken color.

Many housewives lower the window shades that the wall paper may not lose its brilliancy, that the beautiful hues of velvet, satin, and plush tapestry may not be marred by loss in brilliancy and sheen. Bright carpets and rugs are sometimes bought in preference to more delicately tinted ones, because the purchaser knows that the latter will fade quickly if used in a sunny room, and will soon acquire a dull mellow tone. The bright and gay colors and the dull and somber colors are all affected by the sun, but why one should be affected more than another we do not know. Thousands of brilliant and dainty hues catch our eye in the shop and on the street, but not one of them is absolutely permanent; some may last for years, but there is always more or less fading in time.

Sunlight causes many strange, unexplained effects. If the two substances, chlorine and hydrogen, are mixed in a dark room, nothing remarkable occurs any more than though water and milk were mixed, but if a mixture of these substances is exposed to sunlight, a violent explosion occurs and an entirely new substance is formed, a compound entirely different in character from either of its components.

CHAPTER XII 92

By some power not understood by man, the sun is able to form new substances. In the dark, chlorine and hydrogen are simply chlorine and hydrogen; in the sunlight they combine as if by magic into a totally different substance. By the same unexplained power, the sun frequently does just the opposite work; instead of combining two substances to make one new product, the sun may separate or break down some particular substance into its various elements. For example, if the sun's rays fall upon silver chloride, a chemical action immediately begins, and as a result we have two separate substances, chlorine and silver. The sunlight separates silver chloride into its constituents, silver and chlorine.

121. The Magic Wand in Photography. Suppose we coat one side of a glass plate with silver chloride, just as we might put a coat of varnish on a chair. We must be very careful to coat the plate in the dark room,[B] otherwise the sunlight will separate the silver chloride and spoil our plan. Then lay a horseshoe on the plate for good luck, and carry the plate out into the light for a second. The light will separate the silver chloride into chlorine and silver, the latter of which will remain on the plate as a thin film. All of the plate was affected by the sun except the portion protected by the horseshoe which, because it is opaque, would not allow light to pass through and reach the plate. If now the plate is carried back to the dark room and the horseshoe is removed, one would expect to see on the plate an impression of the horseshoe, because the portion protected by the horseshoe would be covered by silver chloride and the exposed unprotected portion would be covered by metallic silver. But we are much disappointed because the plate, when examined ever so carefully, shows not the slightest change in appearance. The change is there, but the unaided eye cannot detect the change. Some chemical, the so-called "developer," must be used to bring out the hidden change and to reveal the image to our unseeing eyes. There are many different developers in use, any one of which will effect the necessary transformation. When the plate has been in the developer for a few seconds, the silver coating gradually darkens, and slowly but surely the image printed by the sun's rays appears. But we must not take this picture into the light, because the silver chloride which was protected by the horseshoe is still present, and would be strongly affected by the first glimmer of light, and, as a result, our entire plate would become similar in character and there would be no

CHAPTER XII 93

contrast to give an image of the horseshoe on the plate.

[Footnote B: That is, a room from which ordinary daylight is excluded.]

But a photograph on glass, which must be carefully shielded from the light and admired only in the dark room, would be neither pleasurable nor practical. If there were some way by which the hitherto unaffected silver chloride could be totally removed, it would be possible to take the plate into any light without fear. To accomplish this, the unchanged silver chloride is got rid of by the process technically called "fixing"; that is, by washing off the unreduced silver chloride with a solution such as sodium thiosulphite, commonly known as hypo. After a bath in the hypo the plate is cleansed in clear running water and left to dry. Such a process gives a clear and permanent picture on the plate.

[Illustration: FIG. 82.--A camera.]

122. The Camera. A camera (Fig. 82) is a light-tight box containing a movable convex lens at one end and a screen at the opposite end. Light from the object to be photographed passes through the lens, falls upon the screen, and forms an image there. If we substitute for the ordinary screen a plate or film coated with silver chloride or any other silver salt, the light which falls upon the sensitive plate and forms an image there will change the silver chloride and produce a hidden image. If the plate is then removed from the camera in the dark, and is treated as described in the preceding Section, the image becomes visible and permanent. In practice some gelatin is mixed with the silver salt, and the mixture is then poured over the plate or film in such a way that a thin, even coating is made. It is the presence of the gelatin that gives plates a yellowish hue. The sensitive plates are left to dry in dark rooms, and when the coating has become absolutely firm and dry, the plates are packed in boxes and sent forth for sale.

Glass plates are heavy and inconvenient to carry, so that celluloid films have almost entirely taken their place, at least for outdoor work.

123. Light and Shade. Let us apply the above process to a real photograph. Suppose we wish to take the photograph of a man sitting in a chair in his library. If the man wore a gray coat, a black tie, and a

white collar, these details must be faithfully represented in the photograph. How can the almost innumerable lights and shades be produced on the plate?

The white collar would send through the lens the most light to the sensitive plate; hence the silver chloride on the plate would be most changed at the place where the lens formed an image of the collar. The gray coat would not send to the lens so much light as the white collar, hence the silver chloride would be less affected by the light from the coat than by that from the collar, and at the place where the lens produced an image of the coat the silver chloride would not be changed so much as where the collar image is. The light from the face would produce a still different effect, since the light from the face is stronger than the light from the gray coat, but less than that from a white collar. The face in the image would show less changed silver chloride than the collar, but more than the coat, because the face is lighter than the coat, but not so light as the collar. Finally, the silver chloride would be least affected by the dark tie. The wall paper in the background would affect the plate according to the brightness of the light which fell directly upon it and which reflected to the camera. When such a plate has been developed and fixed, as described in Section 121, we have the so-called negative (Fig. 83). The collar is very dark, the black tie and gray coat white, and the white tidy very dark.

[Illustration: FIG. 83.--A negative.]

The lighter the object, such as tidy or collar, the more salt is changed, or, in other words, the greater the portion of the silver salt that is affected, and hence the darker the stain on the plate at that particular spot. The plate shows all gradations of intensity--the tidy is dark, the black tie is light. The photograph is true as far as position, form, and expression are concerned, but the actual intensities are just reversed. How this plate can be transformed into a photograph true in every detail will be seen in the following Section.

124. The Perfect Photograph. Bright objects, such as the sky or a white waist, change much of the silver chloride, and hence appear dark on the negative. Dark objects, such as furniture or a black coat, change little of the chloride, and hence appear light on the negative.

CHAPTER XII 95

To obtain a true photograph, the negative is placed on a piece of sensitive photographic paper, or paper coated with a silver salt in the same manner as the plate and films. The combination is exposed to the light. The dark portions of the negative will act as obstructions to the passage of light, and but little light will pass through that part of the negative to the photographic paper, and consequently but little of the silver salt on the paper will be changed. On the other hand, the light portion of the negative will allow free and easy passage of the light rays, which will fall upon the photographic paper and will change much more of the silver. Thus it is that dark places in the negative produce light places in the positive or real photograph (Fig. 84), and that light places in the negative produce dark places in the positive; all intermediate grades are likewise represented with their proper gradations of intensity.

[Illustration: FIG. 84.--A positive or true photograph.]

If properly treated, a negative remains good for years, and will serve for an indefinite number of positives or true photographs.

125. Light and Disease. The far-reaching effect which light has upon some inanimate objects, such as photographic films and clothes, leads us to inquire into the relation which exists between light and living things. We know from daily observation that plants must have light in order to thrive and grow. A healthy plant brought into a dark room soon loses its vigor and freshness, and becomes yellow and drooping. Plants do not all agree as to the amount of light they require, for some, like the violet and the arbutus, grow best in moderate light, while others, like the willows, need the strong, full beams of the sun. But nearly all common plants, whatever they are, sicken and die if deprived of sunlight for a long time. This is likewise true in the animal world. During long transportation, animals are sometimes necessarily confined in dark cars, with the result that many deaths occur, even though the car is well aired and ventilated and the food supply good. Light and fresh air put color into pale cheeks, just as light and air transform sickly, yellowish plants into hardy green ones. Plenty of fresh air, light, and pure water are the watchwords against disease.

CHAPTER XII

[Illustration: FIG. 85--Stems and leaves of oxalis growing toward the light.]

In addition to the plants and animals which we see, there are many strange unseen ones floating in the atmosphere around us, lying in the dust of corner and closet, growing in the water we drink, and thronging decayed vegetable and animal matter. Everyone knows that mildew and vermin do damage in the home and in the field, but very few understand that, in addition to these visible enemies of man, there are swarms of invisible plants and animals some of which do far more damage, both directly and indirectly, than the seen and familiar enemies. All such very small plants and animals are known as *microorganisms*.

Not all microörganisms are harmful; some are our friends and are as helpful to us as are cultivated plants and domesticated animals. Among the most important of the microörganisms are bacteria, which include among their number both friend and foe. In the household, bacteria are a fruitful source of trouble, but some of them are distinctly friends. The delicate flavor of butter and the sharp but pleasing taste of cheese are produced by bacteria. On the other hand, bacteria are the cause of many of the most dangerous diseases, such as typhoid fever, tuberculosis, influenza, and la grippe.

By careful observation and experimentation it has been shown conclusively that sunlight rapidly kills bacteria, and that it is only in dampness and darkness that bacteria thrive and multiply. Although sunlight is essential to the growth of most plants and animals, it retards and prevents the growth of bacteria. Dirt and dust exposed to the sunlight lose their living bacteria, while in damp cellars and dark corners the bacteria thrive, increasing steadily in number. For this reason our houses should be kept light and airy; blinds should be raised, even if carpets do fade; it is better that carpets and furniture should fade than that disease-producing bacteria should find a permanent abode within our dwellings. Kitchens and pantries in particular should be thoroughly lighted. Bedclothes, rugs, and clothing should be exposed to the sunlight as frequently as possible; there is no better safeguard against bacterial disease than light. In a sick room sunlight is especially valuable, because it not only kills bacteria, but keeps the air dry, and new bacteria cannot get a start in a dry atmosphere.

CHAPTER XIII

COLOR

126. The Rainbow. One of the most beautiful and well-known phenomena in nature is the rainbow, and from time immemorial it has been considered Jehovah's signal to mankind that the storm is over and that the sunshine will remain. Practically everyone knows that a rainbow can be seen only when the sun's rays shine upon a mist of tiny drops of water. It is these tiny drops which by their refraction and their scattering of light produce the rainbow in the heavens.

The exquisite tints of the rainbow can be seen if we look at an object through a prism or chandelier crystal, and a very simple experiment enables us to produce on the wall of a room the exact colors of the rainbow in all their beauty.

[Illustration: FIG. 86.--White light is a mixture of lights of rainbow colors.]

127. How to produce Rainbow Colors. *The Spectrum.* If a beam of sunlight is admitted into a dark room through a narrow opening in the shade, and is allowed to fall upon a prism, as shown in Figure 86, a beautiful band of colors will appear on the opposite wall of the room. The ray of light which entered the room as ordinary sunlight has not only been refracted and bent from its straight path, but it has been spread out into a band of colors similar to those of the rainbow.

Whenever light passes through a prism or lens, it is dispersed or separated into all the colors which it contains, and a band of colors produced in this way is called a spectrum. If we examine such a spectrum we find the following colors in order, each color imperceptibly fading into the next: violet, indigo, blue, green, yellow, orange, red.

CHAPTER XIII

128. **Sunlight or White Light.** White light or sunlight can be dispersed or separated into the primary colors or rainbow hues, as shown in the preceding Section. What seems even more wonderful is that these spectral colors can be recombined so as to make white light.

If a prism B (Fig. 87) exactly similar to A in every way is placed behind A in a reversed position, it will undo the dispersion of A, bending upward the seven different beams in such a way that they emerge together and produce a white spot on the screen. Thus we see, from two simple experiments, that all the colors of the rainbow may be obtained from white light, and that these colors may be in turn recombined to produce white light.

[Illustration: FIG. 87.--Rainbow colors recombined to form white light.]

White light is not a simple light, but is composed of all the colors which appear in the rainbow.

129. **Color.** If a piece of red glass is held in the path of the colored beam of light formed as in Section 127, all the colors on the wall will disappear except the red, and instead of a beautiful spectrum of all colors there will be seen the red color alone. The red glass does not allow the passage through it of any light except red light; all other colors are absorbed by the red glass and do not reach the eye. Only the red ray passes through the red glass, reaches the eye, and produces a sensation of color.

If a piece of blue glass is substituted for the red glass, the blue band remains on the wall, while all the other colors disappear. If both blue and red pieces of glass are held in the path of the beam, so that the light must pass through first one and then the other, the entire spectrum disappears and no color remains. The blue glass absorbs the various rays with the exception of the blue ones, and the red glass will not allow these blue rays to pass through it; hence no light is allowed passage to the eye.

An emerald looks green because it freely transmits green, but absorbs the other colors of which ordinary daylight is composed. A diamond appears white because it allows the passage through it of all the

CHAPTER XIII   99

various rays; this is likewise true of water and window panes.

Stained-glass windows owe their charm and beauty to the presence in the glass of various dyes and pigments which absorb in different amounts some colors from white light and transmit others. These pigments or dyes are added to the glass while it is in the molten state, and the beauty of a stained-glass window depends largely upon the richness and the delicacy of the pigments used.

130. Reflected Light. *Opaque Objects.* In Section 106 we learned that most objects are visible to us because of the light diffusely reflected from them. A white object, such as a sheet of paper, a whitewashed fence, or a table cloth, absorbs little of the light which falls upon it, but reflects nearly all, thus producing the sensation of white. A red carpet absorbs the light rays incident upon it except the red rays, and these it reflects to the eye.

Any substance or object which reflects none of the rays which fall upon it, but absorbs all, appears black; no rays reach the eye, and there is an absence of any color sensation. Coal and tar and soot are good illustrations of objects which absorb all the light which falls upon them.

131. How and Why Colors Change. *Matching Colors.* Most women prefer to shop in the morning and early afternoon when the sunlight illuminates shops and factories, and when gas and electricity do not throw their spell over colors. Practically all people know that ribbons and ties, trimmings and dresses, frequently look different at night from what they do in the daytime. It is not safe to match colors by artificial light; cloth which looks red by night may be almost purple by day. Indeed, the color of an object depends upon the color of the light which falls upon it. Strange sights are seen on the Fourth of July when variously colored fireworks are blazing. The child with a white blouse appears first red, then blue, then green, according as his powders burn red, blue, or green. The face of the child changes from its normal healthy hue to a brilliant red and then to ghastly shades.

Suppose, for example, that a white hat is held at the red end of the spectrum or in any red light. The characteristics of white objects is their ability to reflect *all* the various rays that fall upon them. Here, however, the only light which falls upon the white hat is red light, hence the only light which the hat has to reflect is red light and the hat consequently appears red. Similarly, if a white hat is placed in a blue light, it will reflect all the light which falls upon it, namely, blue light, and will appear blue. If a red hat is held in a red light, it is seen in its proper color. If a red hat is held in a blue light, it appears black; it cannot reflect any of the blue light because that is all absorbed and there is no red light to reflect.

A child wearing a green frock on Independence Day seems at night to be wearing a black frock, if standing near powders burning with red, blue, or violet light.

132. Pure, Simple Colors--Things as they Seem. To the eye white light appears a simple, single color. It reveals its compound nature to us only when passed through a prism, when it shows itself to be compounded of an infinite number of colors which Sir Isaac Newton grouped in seven divisions: violet, indigo, blue, green, yellow, orange, and red.

We naturally ask ourselves whether these colors which compose white light are themselves in turn compound? To answer that question, let us very carefully insert a second prism in the path of the rays which issue from the first prism, carefully barring out the remaining six kinds of rays. If the red light is compound, it will be broken up into its constituent parts and will form a typical spectrum of its own, just as white light did after its passage through a prism. But the red rays pass through the second prism, are refracted, and bent from this course, and no new colors appear, no new spectrum is formed. Evidently a ray of spectrum red is a simple color, not a compound color.

If a similar experiment is made with the remaining spectrum rays, the result is always the same: the individual spectrum colors remain simple, pure colors. _The individual spectrum colors are groups of simple, pure colors._

[Illustration: FIG. 88.--Violet and green give blue. Green, blue, and red give white.]

CHAPTER XIII

133. Colors not as they Seem--Compound Colors. If one half of a cardboard disk (Fig. 88) is painted green, and the other half violet, and the disk is slipped upon a toy top, and spun rapidly, the rotating disk will appear blue; if red and green are used in the same way instead of green and violet, the rotating disk will appear yellow. A combination of red and yellow will give orange. The colors formed in this way do not appear to the eye different from the spectrum colors, but they are actually very different. The spectrum colors, as we saw in the preceding Section, are pure, simple colors, while the colors formed from the rotating disk are in reality compounded of several totally different rays, although in appearance the resulting colors are pure and simple.

If it were not that colors can be compounded, we should be limited in hue and shade to the seven spectral colors; the wealth and beauty of color in nature, art, and commerce would be unknown; the flowers with their thousands of hues would have a poverty of color undreamed of; art would lose its magenta, its lilac, its olive, its lavender, and would have to work its wonders with the spectral colors alone. By compounding various colors in different proportions, new colors can be formed to give freshness and variety. If one third of the rotating disk is painted blue, and the remainder white, the result is lavender; if fifteen parts of white, four parts of red, and one part of blue are arranged on the disk, the result is lilac. Olive is obtained from a combination of two parts green, one part red, and one part black; and the soft rich shades of brown are all due to different mixtures of black, red, orange, or yellow.

134. The Essential Colors. Strange and unexpected facts await us at every turn in science! If the rotating cardboard disk (Fig. 88) is painted one third red, one third green, and one third blue, the resulting color is white. While the mixture of the spectral colors produces white, it is not necessary to have all of the spectral colors in order to obtain white; because a mixture of the following colors alone, red, green, and blue, will give white. Moreover, by the mixture of these three colors in proper proportions, any color of the spectrum, such as yellow or indigo or orange, may be obtained. The three spectral colors, red, green, and blue, are called primary or essential hues, because all known tints of color may be produced by the careful blending of blue, green, and red

CHAPTER XIII 102

in the proper proportions; for example, purple is obtained by the blending of red and blue, and orange by the blending of red and yellow.

135. Color Blindness. The nerve fibers of the eye which carry the sensation of color to the brain are particularly sensitive to the primary colors--red, green, blue. Indeed, all color sensations are produced by the stimulation of three sets of nerves which are sensitive to the primary colors. If one sees purple, it is because the optic nerves sensitive to red and blue (purple equals red plus blue) have carried their separate messages to the brain, and the blending of the two distinct messages in the brain has given the sensation of purple. If a red rose is seen, it is because the optic nerves sensitive to red have been stimulated and have carried the message to the brain.

A snowy field stimulates equally all three sets of optic nerves--the red, the green, and the blue. Lavender, which is one part blue and three parts white, would stimulate all three sets of nerves, but with a maximum of stimulation for the blue. Equal stimulation of the three sets would give the impression of white.

A color-blind person has some defect in one or more of the three sets of nerves which carry the color message to the brain. Suppose the nerve fibers responsible for carrying the red are totally defective. If such a person views a yellow flower, he will see it as a green flower. Yellow contains both red and green, and hence both the red and green nerve fibers should be stimulated, but the red nerve fibers are defective and do not respond, the green nerve fibers alone being stimulated, and the brain therefore interprets green.

A well-known author gives an amusing incident of a dinner party, at which the host offered stewed tomato for apple sauce. What color nerves were defective in the case of the host?

In some employments color blindness in an employee would be fatal to many lives. Engineers and pilots govern the direction and speed of trains and boats largely by the colored signals which flash out in the night's darkness or move in the day's bright light, and any mistake in the reading of color signals would imperil the lives of travelers. For this reason a rigid test in color is given to all persons seeking such employment, and the ability to match ribbons and yarns of all ordinary hues is an unvarying requirement for efficiency.

CHAPTER XIV

HEAT AND LIGHT AS COMPANIONS

"The night has a thousand eyes, And the day but one; Yet the light of the bright world dies With the dying sun."

136. Most bodies which glow and give out light are hot; the stove which glows with a warm red is hot and fiery; smoldering wood is black and lifeless; glowing coals are far hotter than black ones. The stained-glass window softens and mellows the bright light of the sun, but it also shuts out some of the warmth of the sun's rays; the shady side of the street spares our eyes the intense glare of the sun, but may chill us by the absence of heat. Our illumination, whether it be oil lamp or gas jet or electric light, carries with it heat; indeed, so much heat that we refrain from making a light on a warm summer's night because of the heat which it unavoidably furnishes.

137. Red a Warm Color. We have seen that heat and light usually go hand in hand. In summer we lower the shades and close the blinds in order to keep the house cool, because the exclusion of light means the exclusion of some heat; in winter we open the blinds and raise the shades in order that the sun may stream into the room and flood it with light and warmth. The heat of the sun and the light of the sun seem boon companions.

We can show that when light passes through a prism and is refracted, forming a spectrum, as in Section 127, it is accompanied by heat. If we hold a sensitive thermometer in the violet end of the spectrum so that the violet rays fall upon the bulb, the reading of the mercury will be practically the same as when the thermometer is held in any dark part of the room; if, however, the thermometer is slowly moved toward the red end of the spectrum, a change occurs and the mercury rises slowly

CHAPTER XIV

but steadily, showing that heat rays are present at the red end of the spectrum. This agrees with the popular notion, formed independently of science, which calls the reds the warm colors. Every one of us associates red with warmth; in the summer red is rarely worn, it looks hot; but in winter red is one of the most pleasing colors because of the sense of warmth and cheer it brings.

All light rays are accompanied by a small amount of heat, but the red rays carry the most.

What seems perhaps the most unexpected thing, is that the temperature, as indicated by a sensitive thermometer, continues to rise if the thermometer is moved just beyond the red light of the spectrum. There actually seems to be more heat beyond the red than in the red, but if the thermometer is moved too far away, the temperature again falls. Later we shall see what this means.

138. The Energy of the Sun. It is difficult to tell how much of the energy of the sun is light and how much is heat, but it is easy to determine the combined effect of heat and light.

[Illustration: FIG. 89.--The energy of the sun can be measured in heat units.]

Suppose we allow the sun's rays to fall perpendicularly upon a metal cylinder coated with lampblack and filled with a known quantity of water (Fig. 89); at the expiration of a few hours the temperature of the water will be considerably higher. Lampblack is a good absorber of heat, and it is used as a coating in order that all the light rays which fall upon the cylinder may be absorbed and none lost by reflection.

Light and heat rays fall upon the lampblack, pass through the cylinder, and heat the water. We know that the red light rays have the largest share toward heating the water, because if the cylinder is surrounded by blue glass which absorbs the red rays and prevents their passage into the water, the temperature of the water begins to fall. That the other light rays have a small share would have been clear from the preceding Section.

CHAPTER XIV

All the energy of the sunshine which falls upon the cylinder, both as heat and as light, is absorbed in the form of heat, and the total amount of this energy can be calculated from the increase in the temperature of the water. The energy which heated the water would have passed onward to the surface of the earth if its path had not been obstructed by the cylinder of water; and we can be sure that the energy which entered the water and changed its temperature would ordinarily have heated an equal area of the earth's surface; and from this, we can calculate the energy falling upon the entire surface of the earth during any one day.

Computations based upon this experiment show that the earth receives daily from the sun the equivalent of 341,000,000,000 horse power--an amount inconceivable to the human mind.

Professor Young gives a striking picture of what this energy of the sun could do. A solid column of ice 93,000,000 miles long and 2-1/4 miles in diameter could be melted in a single second if the sun could concentrate its entire power on the ice.

While the amount of energy received daily from the sun by the earth is actually enormous, it is small in comparison with the whole amount given out by the sun to the numerous heavenly bodies which make up the universe. In fact, of the entire outflow of heat and light, the earth receives only one part in two thousand million, and this is a very small portion indeed.

139. How Light and Heat Travel from the Sun to Us. Astronomers tell us that the sun--the chief source of heat and light--is 93,000,000 miles away from us; that is, so far distant that the fastest express train would require about 176 years to reach the sun. How do heat and light travel through this vast abyss of space?

[Illustration: FIG. 90.--Waves formed by a pebble.]

A quiet pool and a pebble will help to make it clear to us. If we throw a pebble into a quiet pool (Fig. 90), waves or ripples form and spread out in all directions, gradually dying out as they become more and more distant from the pebble. It is a strange fact that while we see the ripple moving farther and farther away, the particles of water are themselves not moving outward and away, but are merely bobbing up and down, or are vibrating. If you wish to be sure of this, throw the pebble near a spot where a chip lies quiet on the smooth pond. After the waves form, the chip rides up and down with the water, but does not move outward; if the water itself were moving outward, it would carry the chip with it, but the water has no forward motion, and hence the chip assumes the only motion possessed by the water, that is, an up-and-down motion. Perhaps a more simple illustration is the appearance of a wheat field or a lawn on a windy day; the wind sweeps over the grass, producing in the grass a wave like the water waves of the ocean, but the blades of grass do not move from their accustomed place in the ground, held fast as they are by their roots.

If a pebble is thrown into a quiet pool, it creates ripples or waves which spread outward in all directions, but which soon die out, leaving the pool again placid and undisturbed. If now we could quickly withdraw the pebble from the pool, the water would again be disturbed and waves would form. If the pebble were attached to a string so that it could be dropped into the water and withdrawn at regular intervals, the waves would never have a chance to disappear, because there would always be a regularly timed definite disturbance of the water. Learned men tell us that all hot bodies and all luminous bodies are composed of tiny particles, called molecules, which move unceasingly back and forth with great speed. In Section 95 we saw that the molecules of all substances move unceasingly; their speed, however, is not so great, nor are their motions so regularly timed as are those of the heat-giving and the light-giving particles. As the particles of the hot and luminous bodies vibrate with great speed and force they violently disturb the medium around them, and produce a series of waves similar to those produced in the water by the pebble. If, however, a pebble is thrown into the water very gently, the disturbance is slight, sometimes too slight to throw the water into waves; in the same way objects whose molecules are in a state of gentle motion do not produce light.

The particles of heat-giving and light-giving bodies are in a state of rapid vibration, and thereby disturb the surrounding medium, which transmits or conveys the disturbance to the earth or to other objects by a train of waves. When these waves reach their destination, the sensation of light or heat is produced.

We see the water waves, but we can never see with the eye the heat and light waves which roll in to us from that far-distant source, the sun. We can be sure of them only through their effect on our bodies, and by the visible work they do.

140. How Heat and Light Differ. If heat and light are alike due to the regular, rapid motion of the particles of a body, and are similarly conveyed by waves, how is it, then, that heat and light are apparently so different?

Light and heat differ as much as the short, choppy waves of the ocean and the slow, long swell of the ocean, but not more so. The sailor handles his boat in one way in a choppy sea and in a different way in a rolling sea, for he knows that these two kinds of waves act dissimilarly. The long, slow swell of the ocean would correspond with the longer, slower waves which travel out from the sun, and which on reaching us are interpreted as heat. The shorter, more frequent waves of the ocean would typify the short, rapid waves which leave the sun, and which on reaching us are interpreted as light.

CHAPTER XV

ARTIFICIAL LIGHTING

141. We seldom consider what life would be without our wonderful methods of illumination which turn night into day, and prolong the hours of work and pleasure. Yet it was not until the nineteenth century that the marvelous change was made from the short-lived candle to the more enduring oil lamp. Before the coming of the lamp, even in large cities like Paris, the only artificial light to guide the belated traveler at night was the candle required to be kept burning in an occasional window.

With the invention of the kerosene lamp came more efficient lighting of home and street, and with the advent of gas and electricity came a

CHAPTER XV

light so effective that the hours of business, manufacture, and pleasure could be extended far beyond the setting of the sun.

The production of light by candle, oil, and gas will be considered in the following paragraphs, while illumination by electricity will be reserved for a later Chapter.

142. The Candle. Candles were originally made by dipping a wick into melting tallow, withdrawing it, allowing the adhered tallow to harden, and repeating the dipping until a satisfactory thickness was obtained. The more modern method consists in pouring a fatty preparation into a mold, at the center of which a wick has been placed.

The wick, when lighted, burns for a brief interval with a faint, uncertain light; almost immediately, however, the intensity of the light increases and the illumination remains good as long as the candle lasts. The heat of the burning tallow melts more of the tallow near it, and this liquid fat is quickly sucked up into the burning wick. The heat of the flame is sufficient to change most of this liquid into a gas, that is, to vaporize the liquid, and furthermore to set fire to the gas thus formed. These heated gases burn with a bright yellow flame.

143. The Oil Lamp. The simple candle of our ancestors was now replaced by the oil lamp, which gave a brighter, steadier, and more permanent illumination. The principle of the lamp is similar to that of the candle, except that the wick is saturated with kerosene or oil rather than with fat. The heat from the burning wick is sufficient to change the oil into a gas and then to set fire to the gas. By placing a chimney over the burning wick, a constant and uniform draught of air is maintained around the blazing gases, and hence a steady, unflickering light is obtained. Gases and carbon particles are set free by the burning wick. In order that the gases may burn and the solid particle glow, a plentiful supply of oxygen is necessary. If the quantity of air is insufficient, the carbon particles remain unburned and form soot. A lamp "smokes" when the air which reaches the wick is insufficient to burn the rapidly formed carbon particles; this explains the danger of turning a lamp wick too high and producing more carbon particles than can be oxidized by the air admitted through the lamp chimney.

CHAPTER XV   109

One great disadvantage of oil lamps and oil stoves is that they cannot be carried safely from place to place. It is almost impossible to carry a lamp without spilling the oil. The flame soon spreads from the wick to the overflowing oil and in consequence the lamp blazes and an explosion may result. Candles, on the other hand, are safe from explosion; the dripping grease is unpleasant but not dangerous.

The illumination from a shaded oil lamp is soft and agreeable, but the trimming of the wicks, the refilling of bowls, and the cleaning of chimneys require time and labor. For this reason, the introduction of gas met with widespread success. The illumination from an ordinary gas jet is stronger than that from an ordinary lamp, and the stronger illumination added to the greater convenience has made gas a very popular source of light.

144. Gas Burners and Gas Mantles. For a long time, the only gas flame used was that in which the luminosity resulted in heating particles of carbon to incandescence. Recently, however, that has been widely replaced by use of a Bunsen flame upon an incandescent mantle, such as the Welsbach. The principle of the incandescent mantle is very simple. When certain substances, such as thorium and cerium, are heated, they do not melt or vaporize, but glow with an intense bright light. Welsbach made use of this fact to secure a burner in which the illumination depends upon the glowing of an incandescent, solid mantle, rather than upon the blazing of a burning gas. He made a cylindrical mantle of thin fabric, and then soaked it in a solution of thorium and cerium until it became saturated with the chemical. The mantle thus impregnated with thorium and cerium is placed on the gas jet, but before the gas is turned on, a lighted match is held to the mantle in order to burn away the thin fabric. After the fabric has been burned away, there remains a coarse gauze mantle of the desired chemicals. If now the gas cock is opened, the escaping gas is ignited, the heat of the flame will raise the mantle to incandescence and will produce a brilliant light. A very small amount of burning gas is sufficient to raise the mantle to incandescence, and hence, by the use of a mantle, intense light is secured at little cost. The mantle saves us gas, because the cock is usually "turned on full" whether we use a plain burner or a mantle burner. But, nevertheless, gas is saved, because when the mantle is adjusted to the gas jet, the

pressure of the gas is lessened by a mechanical device and hence less gas escapes and burns. By actual experiment, it has been found that an ordinary burner consumes about five times as much gas per candle power as the best incandescent burner, and hence is about five times as expensive. One objection to the mantles is their tendency to break. But if the mantles are carefully adjusted on the burner and are not roughly jarred in use, they last many months; and since the best quality cost only twenty-five cents, the expense of renewing the mantles is slight.

145. Gas for Cooking. If a cold object is held in the bright flame of an ordinary gas jet, it becomes covered with soot, or particles of unburned carbon. Although the flame is surrounded by air, the central portion of it does not receive sufficient oxygen to burn up the numerous carbon particles constantly thrown off by the burning gas, and hence many carbon particles remain in the flame as glowing, incandescent masses. That some unburned carbon is present in a flame is shown by the fact that whenever a cold object is held in the flame, it becomes "smoked" or covered with soot. If enough air were supplied to the flame to burn up the carbon as fast as it was set free, there would be no deposition of soot on objects held over the flame or in it, because the carbon would be transformed into gaseous matter.

Unburned carbon would be objectionable in cooking stoves where utensils are constantly in contact with the flame, and for this reason cooking stoves are provided with an arrangement by means of which additional air is supplied to the burning gas in quantities adequate to insure complete combustion of the rapidly formed carbon particles. An opening is made in the tube through which gas passes to the burner, and as the gas moves past this opening, it carries with it a draft of air. These openings are visible on all gas stoves, and should be kept clean and free of clogging, in order to insure complete combustion. So long as the supply of air is sufficient, the flame burns with a dull blue color, but when the supply falls below that needed for complete burning of the carbon, the blue color disappears, and a yellow flame takes its place, and with the yellow flame the deposition of soot is inevitable.

146. By-products of Coal Gas. Many important products besides illuminating gas are obtained from the distillation of soft coal. Ammonia

CHAPTER XV

is made from the liquids which collect in the condensers; anilin, the source of exquisite dyes, is made from the thick, tarry distillate, and coke is the residue left in the clay retorts. The coal tar yields not only anilin, but also carbolic acid and naphthalene, both of which are commercially valuable, the former as a widely used disinfectant, and the latter as a popular moth preventive.

From a ton of good gas-producing coal can be obtained about 10,000 cubic feet of illuminating gas, and as by-products 6 pounds of ammonia, 12 gallons of coal tar, and 1300 pounds of coke.

147. Natural Gas. Animal and vegetable matter buried in the depth of the earth sometimes undergoes natural distillation, and as a result gas is formed. The gas produced in this way is called natural gas. It is a cheap source of illumination, but is found in relatively few localities and only in limited quantity.

148. Acetylene. In 1892 it was discovered that lime and coal fused together in the intense heat of the electric furnace formed a crystalline, metallic-looking substance called calcium carbide. As a result of that discovery, this substance was soon made on a large scale and sold at a moderate price. The cheapness of calcium carbide has made it possible for the isolated farmhouse to discard oil lamps and to have a private gas system. When the hard, gray crystals of calcium carbide are put in water, they give off acetylene, a colorless gas which burns with a brilliant white flame. If bits of calcium carbide are dropped into a test tube containing water, bubbles of gas will be seen to form and escape into the air, and the escaping gas may be ignited by a burning match held near the mouth of the test tube. When chemical action between the water and carbide has ceased, and gas bubbles have stopped forming, slaked lime is all that is left of the dark gray crystals which were put into the water.

When calcium carbide is used as a source of illumination, the crystals are mechanically dropped into a tank containing water, and the gas generated is automatically collected in a small sliding tank, whence it passes through pipes to the various rooms. The slaked lime, formed while the gas was generated, collects at the bottom of the tanks and is removed from time to time.

The cost of an acetylene generator is about $50 for a small house, and the cost of maintenance is not more than that of lamps. The generator does not require filling oftener than once a week, and the labor is less than that required for oil lamps. In a house in which there were twenty burners, the tanks were filled with water and carbide but once a fortnight. Acetylene is seldom used in large cities, but it is very widely used in small communities and is particularly convenient in more or less remote summer residences.

Electric Lights. The most recent and the most convenient lighting is that obtained by electricity. A fine, hairlike filament within a glass bulb is raised to incandescence by the heat of an electric current. This form of illumination will be considered in connection with electricity.

CHAPTER XVI

MAN'S WAY OF HELPING HIMSELF

149. Labor-saving Devices. To primitive man belonged more especially the arduous tasks of the out-of-door life: the clearing of paths through the wilderness; the hauling of material; the breaking up of the hard soil of barren fields into soft loam ready to receive the seed; the harvesting of the ripe grain, etc.

[Illustration: FIG. 91.--Prying a stone out of the ground.]

The more intelligent races among men soon learned to help themselves in these tasks. For example, our ancestors in the field soon learned to pry stones out of the ground (Fig. 91) rather than to undertake the almost impossible task of lifting them out of the earth in which they were embedded; to swing fallen trees away from a path by means of rope attached to one end rather than to attempt to remove them single-handed; to pitch hay rather than to lift it; to clear a field with a rake rather than with the hands; to carry heavy loads in wheelbarrows (Fig. 92) rather than on the shoulders; to roll barrels up a plank (Fig. 93) and to raise weights by ropes. In every case, whether in the lifting of stones, or the felling of trees, or the transportation of

CHAPTER XVI

heavy weights, or the digging of the ground, man used his brain in the invention of mechanical devices which would relieve muscular strain and lighten physical labor.

If all mankind had depended upon physical strength only, the world to-day would be in the condition prevalent in parts of Africa, Asia, and South America, where the natives loosen the soil with their hands or with crude implements (Fig. 94), and transport huge weights on their shoulders and heads.

[Illustration: FIG. 92.--The wheelbarrow lightens labor.]

Any mechanical device (Figs. 95 and 96), whereby man's work can be more conveniently done, is called a machine; the machine itself never does any work--it merely enables man to use his own efforts to better advantage.

[Illustration: FIG. 93.--Rolling barrels up a plank.]

150. When do we Work? Whenever, as a result of effort or force, an object is moved, work is done. If you lift a knapsack from the floor to the table, you do work because you use force and move the knapsack through a distance equal to the height of the table. If the knapsack were twice as heavy, you would exert twice as much force to raise it to the same height, and hence you would do double the work. If you raised the knapsack twice the distance,--say to your shoulders instead of to the level of the table,--you would do twice the work, because while you would exert the same force you would continue it through double the distance.

[Illustration: FIG. 94.--Crude method of farming.]

Lifting heavy weights through great distances is not the only way in which work is done. Painting, chopping wood, hammering, plowing, washing, scrubbing, sewing, are all forms of work. In painting, the moving brush spreads paint over a surface; in chopping wood, the descending ax cleaves the wood asunder; in scrubbing, the wet mop rubbed over the floor carries dirt away; in every conceivable form of work, force and motion occur.

CHAPTER XVI

A man does work when he walks, a woman does work when she rocks in a chair--although here the work is less than in walking. On a windy day the work done in walking is greater than normal. The wind resists our progress, and we must exert more force in order to cover the same distance. Walking through a plowed or rough field is much more tiring than to walk on a smooth road, because, while the distance covered may be the same, the effort put forth is greater, and hence more work is done. Always the greater the resistance encountered, the greater the force required, and hence the greater the work done.

The work done by a boy who raises a 5-pound knapsack to his shoulder would be 5x4, or 20, providing his shoulders were 4 feet from the ground.

The amount of work done depends upon the force used and the distance covered (sometimes called displacement), and hence we can say that

Work = force multiplied by distance, or $W = f \times d$.

151. Machines. A glance into our machine shops, our factories, and even our homes shows how widespread is the use of complex machinery. But all machines, however complicated in appearance, are in reality but modifications and combinations of one or more of four simple machines devised long ago by our remote ancestors. These simple devices are known to-day, as (1) the lever, represented by a crowbar, a pitchfork; (2) the inclined plane, represented by the plank upon which barrels are rolled into a wagon; (3) the pulley, represented by almost any contrivance for the raising of furniture to upper stories; (4) the wheel and axle, represented by cogwheels and coffee grinders.

[Illustration: FIG. 95.--Primitive method of grinding corn.]

Suppose a 600-pound bowlder which is embedded in the ground is needed for the tower of a building. The problem of the builder is to get the heavy bowlder out of the ground, to load it on a wagon for transportation, and finally to raise it to the tower. Obviously, he cannot do this alone; the greatest amount of force of which he is capable would not suffice to accomplish any one of these tasks. How then does

CHAPTER XVI

he help himself and perform the impossible? Simply, by the use of some of the machine types mentioned above, illustrations of which are known in a general way to every schoolboy. The very knife with which a stick is whittled is a machine.

[Illustration: FIG. 96.--Separating rice grains by flailing.]

[Illustration: FIG. 97.--The principle of the lever.]

152. The Lever. Balance a foot rule, containing a hole at its middle point F, as shown in Figure 97. If now a weight of 1 pound is suspended from the bar at some point, say 12, the balance is disturbed, and the bar swings about the point F as a center. The balance can be regained by suspending an equivalent weight at the opposite end of the bar, or by applying a 2-pound weight at a point 3 inches to the left of F. In the latter case a force of 1 pound actually balances a force of 2 pounds, but the 1-pound weight is twice as far from the point of suspension as is the 2-pound weight. The small weight makes up in distance what it lacks in magnitude.

Such an arrangement of a rod or bar is called a lever. In any form of lever there are only three things to be considered: the point where the weight rests, the point where the force acts, and the point called the fulcrum about which the rod rotates.

The distance from the force to the fulcrum is called the force arm. The distance from the weight to the fulcrum is called the weight arm; and it is a law of levers, as well as of all other machines, that the force multiplied by the length of the force arm must equal the weight multiplied by the length of the weight arm.

Force × force arm = weight × weight arm.

A force of 1 pound at a distance of 6, or with a force arm 6, will balance a weight of 2 pounds with a weight arm 3; that is,

1 × 6 = 2 × 3.

CHAPTER XVI

Similarly a force of 10 pounds may be made to sustain a weight of 100 pounds, providing the force arm is 10 times longer than the weight arm; and a force arm of 800 pounds, at a distance of 10 feet from the fulcrum, may be made to sustain a weight of 8000 pounds, providing the weight is 1 foot from the fulcrum.

153. Applications of the Lever. By means of a lever, a 600-pound bowlder can be easily pried out of the ground. Let the lever, any strong metal bar, be supported on a stone which serves as fulcrum; then if a man exerts his force at the end of the rod somewhat as in Figure 91 (p. 154), the force arm will be the distance from the stone or fulcrum to the end of the bar, and the weight arm will be the distance from the fulcrum to the bowlder itself. The man pushes down with a force of 100 pounds, but with that amount succeeds in prying up the 600-pound bowlder. If, however, you look carefully, you will see that the force arm is 6 times as long as the weight arm, so that the smaller force is compensated for by the greater distance through which it acts.

At first sight it seems as though the man's work were done for him by the machine. But this is not so. The man must lower his end of the lever 3 feet in order to raise the bowlder 6 inches out of the ground. He does not at any time exert a large force, but he accomplishes his purpose by exerting a small force continuously through a correspondingly greater distance. He finds it easier to exert a force of 100 pounds continuously until his end has moved 3 feet rather than to exert a force of 600 pounds on the bowlder and move it 6 inches.

By the time the stone has been raised the man has done as much work as though the stone had been raised directly, but his inability to put forth sufficient muscular force to raise the bowlder directly would have rendered impossible a result which was easily accomplished when through the medium of the lever he could extend his small force through greater distance.

154. The Wheelbarrow as a Lever. The principle of the lever is always the same; but the relative position of the important points may vary. For example, the fulcrum is sometimes at one end, the force at the opposite end, and the weight to be lifted between them.

CHAPTER XVI

[Illustration: FIG. 98.--A slightly different form of lever.]

Suspend a stick with a hole at its center as in Figure 98, and hang a 4-pound weight at a distance of 1 foot from the fulcrum, supporting the load by means of a spring balance 2 feet from the fulcrum. The pointer on the spring balance shows that the force required to balance the 4-pound load is but 2 pounds.

The force is 2 feet from the fulcrum, and the weight (4) is 1 foot from the fulcrum, so that

Force × distance = Weight × distance, or 2 × 2 = 4 × 1.

Move the 4-pound weight so that it is very near the fulcrum, say but 6 inches from it; then the spring balance registers a force only one fourth as great as the weight which it suspends. In other words a force of 1 at a distance of 24 inches (2 feet) is equivalent to a force of 4 at a distance of 6 inches.

[Illustration: FIG. 99.--The wheelbarrow lightened labor.]

One of the most useful levers of this type is the wheelbarrow (Fig. 99). The fulcrum is at the wheel, the force is at the handles, the weight is on the wheelbarrow. If the load is halfway from the fulcrum to the man's hands, the man will have to lift with a force equal to one half the load. If the load is one fourth as far from the fulcrum as the man's hands, he will need to lift with a force only one fourth as great as that of the load.

[Illustration: FIG. 100.--A modified wheelbarrow.]

This shows that in loading a wheelbarrow, it is important to arrange the load as near to the wheel as possible.

[Illustration: FIG. 101.--The nutcracker is a lever.]

The nutcracker (Fig. 101) is an illustration of a double lever of the wheelbarrow kind; the nearer the nut is to the fulcrum, the easier the cracking.

CHAPTER XVI

[Illustration: FIG. 102.--The hand exerts a small force over a long distance and draws out a nail.]

Hammers (Fig. 102), tack-lifters, scissors, forceps, are important levers, and if you will notice how many different levers (fig. 103) are used by all classes of men, you will understand how valuable a machine this simple device is.

155. The Inclined Plane. A man wishes to load the 600-pound bowlder on a wagon, and proceeds to do it by means of a plank, as in Figure 93. Such an arrangement is called an inclined plane.

The advantage of an inclined plane can be seen by the following experiment. Select a smooth board 4 feet long and prop it so that the end A (Fig. 104) is 1 foot above the level of the table; the length of the incline is then 4 times as great as its height. Fasten a metal roller to a spring balance and observe its weight. Then pull the roller uniformly upward along the plank and notice what the pull is on the balance, being careful always to hold the balance parallel to the incline.

When the roller is raised along the incline, the balance registers a pull only one fourth as great as the actual weight of the roller. That is, when the roller weighs 12, a force of 3 suffices to raise it to the height A along the incline; but the smaller force must be applied throughout the entire length of the incline. In many cases, it is preferable to exert a force of 30 pounds, for example, over the distance CA than a force of 120 pounds over the shorter distance BA.

[Illustration: FIG. 103.--Primitive man tried to lighten his task by balancing his burden.]

Prop the board so that the end A is 2 feet above the table level; that is, arrange the inclined plane in such a way that its length is twice as great as its height. In that case the steady pull on the balance will be one half the weight of the roller; or a force of 6 pounds will suffice to raise the 12-pound roller.

[Illustration: FIG. 104.--Less force is required to raise the roller along the incline than to raise it to A directly.]

CHAPTER XVI

The steeper the incline, the more force necessary to raise a weight; whereas if the incline is small, the necessary lifting force is greatly reduced. On an inclined plane whose length is ten times its height, the lifting force is reduced to one tenth the weight of the load. The advantage of an incline depends upon the relative length and height, or is equal to the ratio of the length to the height.

156. Application. By the use of an inclined plank a strong man can load the 600-pound bowlder on a wagon. Suppose the floor of the wagon is 2 feet above the ground, then if a 6-foot plank is used, 200 pounds of force will suffice to raise the bowlder; but the man will have to push with this force against the bowlder while it moves over the entire length of the plank.

Since work is equal to force multiplied by distance, the man has done work represented by 200 × 6, or 1200. This is exactly the amount of work which would have been necessary to raise the bowlder directly. A man of even enormous strength could not lift such a weight (600 lb.) even an inch directly, but a strong man can furnish the smaller force (200) over a distance of 6 feet; hence, while the machine does not lessen the total amount of work required of a man, it creates a new distribution of work and makes possible, and even easy, results which otherwise would be impossible by human agency.

157. Railroads and Highways. The problem of the incline is an important one to engineers who have under their direction the construction of our highways and the laying of our railroad tracks. It requires tremendous force to pull a load up grade, and most of us are familiar with the struggling horse and the puffing locomotive. For this reason engineers, wherever possible, level down the steep places, and reduce the strain as far as possible.

[Illustration: FIG. 105.--A well-graded railroad bed.]

The slope of the road is called its grade, and the grade itself is simply the number of feet the hill rises per mile. A road a mile long (5280 feet) has a grade of 132 if the crest of the hill is 132 feet above the level at which the road started.

CHAPTER XVI

[Illustration: FIG. 106.--A long, gradual ascent is better than a shorter, steeper one.]

In such an incline, the ratio of length to height is 5280 ÷ 132, or 40; and hence in order to pull a train of cars to the summit, the engine would need to exert a continuous pull equal to one fortieth of the combined weight of the train.

If, on the other hand, the ascent had been gradual, so that the grade was 66 feet per mile, a pull from the engine of one eightieth of the combined weight would have sufficed to land the train of cars at the crest of the grade.

Because of these facts, engineers spend great sums in grading down railroad beds, making them as nearly level as possible. In mountainous regions, the topography of the land prevents the elimination of all steep grades, but nevertheless the attempt is always made to follow the easiest grades.

158. The Wedge. If an inclined plane is pushed underneath or within an object, it serves as a wedge. Usually a wedge consists of two inclined planes (Fig. 107).

[Illustration: FIG. 107.--By means of a wedge, the stump is split.]

A chisel and an ax are illustrations of wedges. Perhaps the most universal form of a wedge is our common pin. Can you explain how this is a wedge?

159. The Screw. Another valuable and indispensable form of the inclined plane is the screw. This consists of a metal rod around which passes a ridge, and Figure 108 shows clearly that a screw is simply a rod around which (in effect) an inclined plane has been wrapped.

[Illustration: FIG. 108--A screw as a simple machine.]

The ridge encircling the screw is called the thread, and the distance between two successive threads is called the pitch. It is easy to see that the closer the threads and the smaller the pitch, the greater the

CHAPTER XVI   121

advantage of the screw, and hence the less force needed in overcoming resistance. A corkscrew is a familiar illustration of the screw.

160. Pulleys. The pulley, another of the machines, is merely a grooved wheel around which a cord passes. It is sometimes more convenient to move a load in one direction rather than in another, and the pulley in its simplest form enables us to do this. In order to raise a flag to the top of a mast, it is not necessary to climb the mast, and so pull up the flag; the same result is accomplished much more easily by attaching the flag to a movable string, somewhat as in Figure 109, and pulling from below. As the string is pulled down, the flag rises and ultimately reaches the desired position.

If we employ a stationary pulley, as in Figure 109, we do not change the force, because the force required to balance the load is as large as the load itself. The only advantage is that a force in one direction may be used to produce motion in another direction. Such a pulley is known as a fixed pulley.

[Illustration: FIG. 109.--By means of a pulley, a force in one direction produces motion in the opposite direction.]

161. Movable Pulleys. By the use of a movable pulley, we are able to support a weight by a force equal to only one half the load. In Figure 109, the downward pull of the weight and the downward pull of the hand are equal; in Figure 110, the spring balance supports only one half the entire load, the remaining half being borne by the hook to which the string is attached. The weight is divided equally between the two parts of the string which passes around the pulley, so that each strand bears only one half of the burden.

We have seen in our study of the lever and the inclined plane that an increase in force is always accompanied by a decrease in distance, and in the case of the pulley we naturally look for a similar result. If you raise the balance (Fig. 110) 12 feet, you will find that the weight rises only 6 feet; if you raise the balance 24 inches, you will find that the weight rises 12 inches. You must exercise a force of 100 pounds over 12 feet of space in order to raise a weight of 200 pounds a distance of

CHAPTER XVI

6 feet. When we raise 100 pounds through 12 feet or 200 pounds through 6 feet the total work done is the same; but the pulley enables those who cannot furnish a force of 200 pounds for the space of 6 feet to accomplish the task by furnishing 100 pounds for the space of 12 feet.

[Illustration: FIG. 110.--A movable pulley lightens labor.]

162. Combination of Pulleys. A combination of pulleys called block and tackle is used where very heavy loads are to be moved. In Figure 111 the upper block of pulleys is fixed, the lower block is movable, and one continuous rope passes around the various pulleys. The load is supported by 6 strands, and each strand bears one sixth of the load. If the hand pulls with a force of 1 pound at P, it can raise a load of 6 pounds at W, but the hand will have to pull downward 6 feet at P in order to raise the load at W 1 foot. If 8 pulleys were used, a force equivalent to one eighth of the load would suffice to move W, but this force would have to be exerted over a distance 8 times as great as that through which W was raised.

[Illustration: FIG. 111.--An effective arrangement of pulleys known as block and tackle.]

163. Practical Application. In our childhood many of us saw with wonder the appearance and disappearance of flags flying at the tops of high masts, but observation soon taught us that the flags were raised by pulleys. In tenements, where there is no yard for the family washing, clothes often appear flapping in mid-air. This seems most marvelous until we learn that the lines are pulled back and forth by pulleys at the window and at a distant support. By means of pulleys, awnings are raised and lowered, and the use of pulleys by furniture movers, etc., is familiar to every wide-awake observer on the streets.

164. Wheel and Axle. The wheel and axle consists of a large wheel and a small axle so fastened that they rotate together.

[Illustration: FIG. 112.--The wheel and axle.]

CHAPTER XVI

When the large wheel makes one revolution, P falls a distance equal to the circumference of the wheel. While P moves downward, W likewise moves, but its motion is upward, and the distance it moves is small, being equal only to the circumference of the small axle. But a small force at P will sustain a larger force at W; if the circumference of the large wheel is 40 inches, and that of the small wheel 10 inches, a load of 100 at W can be sustained by a force of 25 at P. The increase in force of the wheel and axle depends upon the relative size of the two parts, that is, upon the circumference of wheel as compared with circumference of axle, and since the ratio between circumference and radius is constant, the ratio of the wheel and axle combination is the ratio of the long radius to the short radius.

For example, in a wheel and axle of radii 20 and 4, respectively, a given weight at P would balance 5 times as great a load at W.

165. Application. *Windlass, Cogwheels.* In the old-fashioned windlass used in farming districts, the large wheel is replaced by a handle which, when turned, describes a circle. Such an arrangement is equivalent to wheel and axle (Fig. 112); the capstan used on shipboard for raising the anchor has the same principle. The kitchen coffee grinder and the meat chopper are other familiar illustrations.

Cogwheels are modifications of the wheel and axle. Teeth cut in A fit into similar teeth cut in B, and hence rotation of A causes rotation of B. Several revolutions of the smaller wheel, however, are necessary in order to turn the larger wheel through one complete revolution; if the radius of A is one half that of B, two revolutions of A will correspond to one of B; if the radius of A is one third that of B, three revolutions of A will correspond to one of B.

[Illustration: FIG. 113.--Cogwheels.]

Experiment demonstrates that a weight W attached to a cogwheel of radius 3 can be raised by a force P, equal to one third of W applied to a cogwheel of radius 1. There is thus a great increase in force. But the speed with which W is raised is only one third the speed with which the small wheel rotates, or increase in power has been at the decrease of speed.

This is a very common method for raising heavy weights by small force.

Cogwheels can be made to give speed at the decrease of force. A heavy weight W attached to B will in its slow fall cause rapid rotation of A, and hence rapid rise of P. It is true that P, the load raised, will be less than W, the force exerted, but if speed is our aim, this machine serves our purpose admirably.

An extremely important form of wheel and axle is that in which the two wheels are connected by belts as in Figure 114. Rotation of W induces rotation of w, and a small force at W is able to overcome a large force at w. An advantage of the belt connection is that power at one place can be transmitted over a considerable distance and utilized in another place.

[Illustration: FIG. 114.--By means of a belt, motion can be transferred from place to place.]

166. Compound Machines. Out of the few simple machines mentioned in the preceding Sections has developed the complex machinery of to-day. By a combination of screw and lever, for example, we obtain the advantage due to each device, and some compound machines have been made which combine all the various kinds of simple machines, and in this way multiply their mechanical advantage many fold.

A relatively simple complex machine called the crane (Fig. 116) maybe seen almost any day on the street, or wherever heavy weights are being lifted. It is clear that a force applied to turn wheel 1 causes a slower rotation of wheel 3, and a still slower rotation of wheel 4, but as 4 rotates it winds up a chain and slowly raises Q. A very complex machine is that seen in Figure 117.

[Illustration: FIG. 115.--A simple derrick for raising weights.]

[Illustration: FIG. 116.--A traveling crane.]

167. **Measurement of Work.** In Section 150, we learned that the amount of work done depends upon the force exerted, and the distance covered, or that W = force × distance. A man who raises 5 pounds a height of 5 feet does far more work than a man who raises 5 ounces a height of 5 inches, but the product of force by distance is 25 in each case. There is difficulty because we have not selected an arbitrary unit of work. The unit of work chosen and in use in practical affairs is the foot pound, and is defined as the work done when a force of 1 pound acts through a distance of 1 foot. A man who moves 8 pounds through 6 feet does 48 foot pounds of work, while a man who moves 8 ounces (1/2 pound) through 6 inches (1/2 foot) does only one fourth of a foot pound of work.

[Illustration: FIG. 117.--A farm engine putting in a crop.]

168. **The Power or the Speed with which Work is Done.** A man can load a wagon more quickly than a growing boy. The work done by the one is equal to the work done by the other, but the man is more powerful, because the time required for a given task is very important. An engine which hoists a 50-pound weight in 1 second is much more powerful than a man who requires 50 seconds for the same task; hence in estimating the value of a working agent, whether animal or mechanical, we must consider not only the work done, but the speed with which it is done.

The rate at which a machine is able to accomplish a unit of work is called *power*, and the unit of power customarily used is the horse power. Any power which can do 550 foot pounds of work per second is said to be one horse power (H.P.). This unit was chosen by James Watt, the inventor of a steam engine, when he was in need of a unit with which to compare the new source of power, the engine, with his old source of power, the horse. Although called a horse power it is greater than the power of an average horse.

An ordinary man can do one sixth of a horse power. The average locomotive of a railroad has more than 500 H.P., while the engines of an ocean liner may have as high as 70,000 H.P.

CHAPTER XVI

169. Waste Work and Efficient Work. In our study of machines we omitted a factor which in practical cases cannot be ignored, namely, friction. No surface can be made perfectly smooth, and when a barrel rolls over an incline, or a rope passes over a pulley, or a cogwheel turns its neighbor, there is rubbing and slipping and sliding. Motion is thus hindered, and the effective value of the acting force is lessened. In order to secure the desired result it is necessary to apply a force in excess of that calculated. This extra force, which must be supplied if friction is to be counteracted, is in reality waste work.

If the force required by a machine is 150 pounds, while that calculated as necessary is 100 pounds, the loss due to friction is 50 pounds, and the machine, instead of being thoroughly efficient, is only two thirds efficient.

Machinists make every effort to eliminate from a machine the waste due to friction, leveling and grinding to the most perfect smoothness and adjustment every part of the machine. When the machine is in use, friction may be further reduced by the use of lubricating oil. Friction can never be totally eliminated, however, and machines of even the finest construction lose by friction some of their efficiency, while poorly constructed ones lose by friction as much as one half of their efficiency.

170. Man's Strength not Sufficient for Machines. A machine, an inert mass of metal and wood, cannot of itself do any work, but can only distribute the energy which is brought to it. Fortunately it is not necessary that this energy should be contributed by man alone, because the store of energy possessed by him is very small in comparison with the energy required to run locomotives, automobiles, sawmills, etc. Perhaps the greatest value of machines lies in the fact that they enable man to perform work by the use of energy other than his own.

[Illustration: FIG. 118.--Man's strength is not sufficient for heavy work.]

Figure 118 shows one way in which a horse's energy can be utilized in lifting heavy loads. Even the fleeting wind has been harnessed by man, and, as in the windmill, made to work for him (Fig. 119). One

CHAPTER XVII   127

sees dotted over the country windmills large and small, and in Holland, the country of windmills, the landowner who does not possess a windmill is poor indeed.

For generations running water from rivers, streams, and falls has served man by carrying his logs downstream, by turning the wheels of his mill, etc.; and in our own day running water is used as an indirect source of electric lights for street and house, the energy of the falling water serving to rotate the armature of a dynamo (Section 310).

[Illustration: FIG. 119.--The windmill pumps water into the troughs where cattle drink.]

A more constant source of energy is that available from the burning of fuel, such as coal and oil. The former is the source of energy in locomotives, the latter in most automobiles.

In the following Chapter will be given an account of water, wind, and fuel as machine feeders.

CHAPTER XVII

THE POWER BEHIND THE ENGINE

171. Small boys soon learn the power of running water; swimming or rowing downstream is easy, while swimming or rowing against the current is difficult, and the swifter the water, the easier the one and the more difficult the other; the river assists or opposes us as we go with it or against it. The water of a quiet pool or of a gentle stream cannot do work, but water which is plunging over a precipice or dam, or is flowing down steep slopes, may be made to saw wood, grind our corn, light our streets, run our electric cars, etc. A waterfall, or a rapid stream, is a great asset to any community, and for this reason should be carefully guarded. Water power is as great a source of wealth as a coal bed or a gold mine.

CHAPTER XVII   128

The most tremendous waterfall in our country is Niagara Falls, which every minute hurls millions of gallons of water down a 163-foot precipice. The energy possessed by such an enormous quantity of water flowing at such a tremendous speed is almost beyond everyday comprehension, and would suffice to run the engines of many cities far and near. Numerous attempts to buy from the United States the right to utilize some of this apparently wasted energy have been made by various commercial companies. It is fortunate that these negotiations have been largely fruitless, because much deviation of the water for commercial uses and the installation of machinery in the vicinity of the famous falls would greatly detract from the beauty of this world-known scene, and would rob our country of a natural beauty unequaled elsewhere.

[Illustration: FIG. 120.--A mountain stream turns the wheels of the mill.]

172. Water Wheels. In Figure 120 the water of a small but rapid mountain stream is made to rotate a large wheel, which in turn communicates its motion through belts to a distant sawmill or grinder. In more level regions huge dams are built which hold back the water and keep it at a higher level than the wheel; from the dam the water is conveyed in pipes (flumes) to the paddle wheel which it turns. Cogwheels or belts connect the paddle wheel with the factory machinery, so that motion of the paddle wheel insures the running of the machinery.

[Illustration: FIG. 121.--The Pelton water wheel.]

One of the most efficient forms of water wheels is that shown in Figure 121, and called the Pelton wheel. Water issues in a narrow jet similar to that of the ordinary garden hose and strikes with great force against the lower part of the wheel, thereby causing rotation of the wheel. Belts transfer this motion to the machinery of factory or mill.

173. Turbines. The most efficient form of water motor is the turbine, a strong metal wheel shaped somewhat like a pin wheel, inclosed in a heavy metal case.

[Illustration: FIG. 122--A turbine at Niagara Falls.]

CHAPTER XVII

Water is conveyed from a reservoir or dam through a pipe (penstock) to the turbine case, in which is placed the heavy metal turbine wheel (Fig. 122). The force of the water causes rotation of the turbine and of the shaft which is rigidly fastened to it. The water which flows into the turbine case causes rotation of the wheel, escapes from the case through openings, and flows into the tail water.

The power which a turbine can furnish depends upon the quantity of water and the height of the fall, and also upon the turbine wheel itself. One of the largest turbines known has a horse power of about 20,000; that is, it is equivalent, approximately, to 20,000 horses.

174. How much is a Stream Worth? The work which a stream can perform may be easily calculated. Suppose, for example, that 50,000 pounds of water fall over a 22-foot dam every second; the power of such a stream would be 1,100,000 foot pounds per second or 2000 H.P. Naturally, a part of this power would be lost to use by friction within the machinery and by leakage, so that the power of a turbine run by a 2000 H.P. stream would be less than that value.

Of course, the horse power to be obtained from a stream determines the size of the paddle wheel or turbine which can be run by it. It would be possible to construct a turbine so large that the stream would not suffice to turn the wheel; for this reason, the power of a stream is carefully determined before machine construction is begun, and the size of the machinery depends upon the estimates of the water power furnished by expert engineers.

A rough estimate of the volume of a stream may be made by the method described below:--

Suppose we allow a stream of water to flow through a rectangular trough; the speed with which the water flows through the trough can be determined by noting the time required for a chip to float the length of the trough; if the trough is 10 feet long and the time required is 5 seconds, the water has a velocity of 2 feet per second.

[Illustration: FIG. 123.--Estimating the quantity of water which flows through the trough each second.]

CHAPTER XVII

The quantity of water which flows through the trough each second depends upon the dimensions of the trough and the velocity of the water. Suppose the trough is 5 feet wide and 3 feet high, or has a cross section of 15 square feet. If the velocity of the water were 1 foot per second, then 15 cubic feet of water would pass any given point each second, but since the velocity of the water is 2 feet per second, 30 cubic feet will represent the amount of water which will flow by a given point in one second.

175. **Quantity of Water Furnished by a River.** Drive stakes in the river at various places and note the time required for a chip to float from one stake to another. If we know the distance between the stakes and the time required for the chip to float from one stake to another, the velocity of the water can be readily determined.

The width of the stream from bank to bank is easily measured, and the depth is obtained in the ordinary way by sounding; it is necessary to take a number of soundings because the bed of the river is by no means level, and soundings taken at only one level would not give an accurate estimate. If the soundings show the following depths: 30, 25, 20, 32, 28, the average depth could be taken as 30 + 25 + 20 + 32 + 28 ÷ 5, or 27 feet. If, as a result of measuring, the river at a given point in its course is found to be 27 feet deep and 60 feet wide, the area of a cross section at that spot would be 1620 square feet, and if the velocity proved to be 6 feet per second, then the quantity of water passing in any one second would be 1620 × 6, or 9720 cubic feet. By experiment it has been found that 1 cu. ft. of water weighs about 62.5 lb. The weight of the water passing each second would therefore be 62.5 × 9720, or 607,500 lb. If this quantity of water plunges over a 10-ft. dam, it does 607,500 × 10, or 6,075,000 foot pounds of work per second, or 11,045 H.P. Such a stream would be very valuable for the running of machinery.

176. **Windmills.** Those of us who have spent our vacation days in the country know that there is no ready-made water supply there as in the cities, but that as a rule the farmhouses obtain their drinking water from springs and wells. In poorer houses, water is laboriously carried in buckets from the spring or is lifted from the well by the windlass. In more prosperous houses, pumps are installed; this is an improvement

CHAPTER XVII

over the original methods, but the quantity of water consumed by the average family is so great as to make the task of pumping an arduous one.

The average amount of water used per day by one person is 25 gallons. This includes water for drinking, cooking, dish washing, bathing, laundry. For a family of five, therefore, the daily consumption would be 125 gallons; if to this be added the water for a single horse, cow, and pig, the total amount needed will be approximately 150 gallons per day. A strong man can pump that amount from an ordinary well in about one hour, but if the well is deep, more time and strength are required.

The invention of the windmill was a great boon to country folks because it eliminated from their always busy life one task in which labor and time were consumed.

177. The Principle of the Windmill. The toy pin wheel is a windmill in miniature. The wind strikes the sails, and causes rotation; and the stronger the wind blows, the faster will the wheel rotate. In windmills, the sails are of wood or steel, instead of paper, but the principle is identical.

[Illustration: FIG. 124.--The toy pin wheel is a miniature windmill.]

As the wheel rotates, its motion is communicated to a mechanical device which makes use of it to raise and lower a plunger, and hence as long as the wind turns the windmill, water is raised. The water thus raised empties into a large tank, built either in the windmill tower or in the garret of the house, and from the tank the water flows through pipes to the different parts of the house. On very windy days the wheel rotates rapidly, and the tank fills quickly; in order to guard against an overflow from the tank a mechanical device is installed which stops rotation of the wheel when the tank is nearly full. The supply tank is usually large enough to hold a supply of water sufficient for several days, and hence a continuous calm of a day or two does not materially affect the house flow. When once built, a windmill practically takes care of itself, except for oiling, and is an efficient and cheap domestic possession.

CHAPTER XVII

[Illustration: FIG. 125.--The windmill pumps water into the tank.]

178. **Steam as a Working Power.** If a delicate vane is held at an opening from which steam issues, the pressure of the steam will cause rotation of the vane (Fig. 126), and if the vane is connected with a machine, work can be obtained from the steam.

When water is heated in an open vessel, the pressure of its steam is too low to be of practical value, but if on the contrary water is heated in an almost closed vessel, its steam pressure is considerable. If steam at high pressure is directed by nozzles against the blades of a wheel, rapid rotation of the wheel ensues just as it did in Figure 121, although in this case steam pressure replaces water pressure. After the steam has spent itself in turning the turbine, it condenses into water and makes its escape through openings in an inclosing case. In Figure 127 the protecting case is removed, in order that the form of the turbine and the positions of the nozzles may be visible.

[Illustration: FIG. 126.--Steam as a source of power.]

[Illustration: FIG. 127.--Steam turbine with many blades and 4 nozzles.]

A single large turbine wheel may have as many as 800,000 sails or blades, and steam may pour out upon these from many nozzles.

The steam turbine is very much more efficient than its forerunner, the steam engine. The installation of turbines on ocean liners has been accompanied by great increase in speed, and by an almost corresponding decrease in the cost of maintenance.

179. **Steam Engines.** A very simple illustration of the working of a steam engine is given in Figure 128. Steam under pressure enters through the opening F, passes through N, and presses upon the piston M. As a result M moves downward, and thereby induces rotation in the large wheel L.

[Illustration: FIG. 128.--The principle of the steam engine.]

CHAPTER XVII

As *M* falls it drives the air in *D* out through *O* and *P* (the opening *P* is not visible in the diagram).

As soon as this is accomplished, a mechanical device draws up the rod *E*, which in turn closes the opening *N*, and thus prevents the steam from passing into the part of *D* above *M*.

But when the rod *E* is in such a position that *N* is closed, *O* on the other hand is open, and steam rushes through it into *D* and forces up the piston. This up-and-down motion of the piston causes continuous rotation of the wheel *L*. If the fire is hot, steam is formed quickly, and the piston moves rapidly; if the fire is low, steam is formed slowly, and the piston moves less rapidly.

The steam engine as seen on our railroad trains is very complex, and cannot be discussed here; in principle, however, it is identical with that just described. Figure 129 shows a steam harvester at work on a modern farm.

[Illustration: FIG. 129.--Steam harvester at work.]

In both engine and turbine the real source of power is not the steam but the fuel, such as coal or oil, which converts the water into steam.

180. Gas Engines. Automobiles have been largely responsible for the gas engine. To carry coal for fuel and water for steam would be impracticable for most motor cars. Electricity is used in some cars, but the batteries are heavy, expensive, and short-lived, and are not always easily replaceable. For this reason gasoline is extensively used, and in the average automobile the source of power is the force generated by exploding gases.

It was discovered some years ago that if the vapor of gasoline or naphtha was mixed with a definite quantity of air, and a light was applied to the mixture, an explosion would result. Modern science uses the force of such exploding gases for the accomplishment of work, such as running of automobiles and launches.

CHAPTER XVIII

In connection with the gasoline supply is a carburetor or sprayer, from which the cylinder C (Fig. 130) receives a fine mist of gasoline vapor and air. This mixture is ignited by an automatic, electric sparking device, and the explosion of the gases drives the piston P to the right. In the 4-cycle type of gas engines (Fig. 130)--the kind used in automobiles--the four strokes are as follows: 1. The mixture of gasoline and air enters the cylinder as the piston moves to the right. 2. The valves being closed, the mixture is compressed as the piston moves to the left. 3. The electric spark ignites the compressed mixture and drives the piston to the right. 4. The waste gas is expelled as the piston moves to the left. The exhaust valve is then closed, the inlet valve opened, and another cycle of four strokes begins.

[Illustration: FIG. 130.--The gas engine.]

The use of gasoline in launches and automobiles is familiar to many. Not only are launches and automobiles making use of gas power, but the gasoline engine has made it possible to propel aëroplanes through the air.

CHAPTER XVIII

PUMPS AND THEIR VALUE TO MAN

181. "As difficult as for water to run up a hill!" Is there any one who has not heard this saying? And yet most of us accept as a matter of course the stream which gushes from our faucet, or give no thought to the ingenuity which devised a means of forcing water upward through pipes. Despite the fact that water flows naturally down hill, and not up, we find it available in our homes and office buildings, in some of which it ascends to the fiftieth floor; and we see great streams of it directed upon the tops of burning buildings by firemen in the streets below.

In the country, where there are no great central pumping stations, water for the daily need must be raised from wells, and the supply of each household is dependent upon the labor and foresight of its members. The water may be brought to the surface either by

CHAPTER XVIII 135

laboriously raising it, bucket by bucket, or by the less arduous method of pumping. These are the only means possible; even the windmill does not eliminate the necessity for the pump, but merely replaces the energy used by man in working it.

In some parts of our country we have oil beds or wells. But if this underground oil is to be of service to man, it must be brought to the surface, and this is accomplished, as in the case of water, by the use of pumps.

An old tin can or a sponge may serve to bale out water from a leaking rowboat, but such a crude device would be absurd if employed on our huge vessels of war and commerce. Here a rent in the ship's side would mean inevitable loss were it not possible to rid the ship of the inflowing water by the action of strong pumps.

Another and very different use to which pumps are put is seen in the compression of gases. Air is forced into the tires of bicycles and automobiles until they become sufficiently inflated to insure comfort in riding. Some present-day systems of artificial refrigeration (Section 93) could not exist without the aid of compressed gases.

Compressed air has played a very important role in mining, being sent into poorly ventilated mines to improve the condition of the air, and to supply to the miners the oxygen necessary for respiration. Divers and men who work under water carry on their backs a tank of compressed air, and take from it, at will, the amount required.

There are many forms of pumps, and they serve widely different purposes, being essential to the operation of many industrial undertakings. In the following Sections some of these forms will be studied.

[Illustration: FIG. 131.--Carrying water home from the spring.]

182. The Air as Man's Servant. Long before man harnessed water for turbines, or steam for engines, he made the air serve his purpose, and by means of it raised water from hidden underground depths to the surface of the earth; likewise, by means of it, he raised to his dwelling

CHAPTER XVIII
136

on the hillside water from the stream in the valley below. Those who live in cities where running water is always present in the home cannot realize the hardship of the days when this "ready-made" supply did not exist, but when man laboriously carried to his dwelling, from distant spring and stream, the water necessary for the daily need.

What are the characteristics of the air which have enabled man to accomplish these feats? They are well known to us and may be briefly stated as follows:--

(1) Air has weight, and 1 cubic foot of air, at atmospheric pressure, weighs 1-1/4 ounces.

(2) The air around us presses with a force of about 15 pounds upon every square inch of surface that it touches.

(3) Air is elastic; it can be compressed, as in the balloon or bicycle tire, but it expands immediately when pressure is reduced. As it expands and occupies more space, its pressure falls and it exerts less force against the matter with which it comes in contact. If, for example, 1 cubic foot of air is allowed to expand and occupy 2 cubic feet of space, the pressure which it exerts is reduced one half. When air is compressed, its pressure increases, and it exerts a greater force against the matter with which it comes in contact. If 2 cubic feet of air are compressed to 1 cubic foot, the pressure of the compressed air is doubled. (See Section 89.)

[Illustration: FIG. 132.--The atmosphere pressing downward on *a* pushes water after the rising piston *b*.]

183. The Common Pump or Lifting Pump. Place a tube containing a close-fitting piston in a vessel of water, as shown in Figure 132. Then raise the piston with the hand and notice that the water rises in the piston tube. The rise of water in the piston tube is similar to the raising of lemonade through a straw (Section 77). The atmosphere presses with a force of 15 pounds upon every square inch of water in the large vessel, and forces some of it into the space left vacant by the retreating piston. The common pump works in a similar manner. It consists of a piston or plunger which moves back and forth in an

CHAPTER XVIII

air-tight cylinder, and contains an outward opening valve through which water and air can pass. From the bottom of the cylinder a tube runs down into the well or reservoir, and water from the well has access to the cylinder through another outward-moving valve. In practice the tube is known as the suction pipe, and its valve as the suction valve.

In order to understand the action of a pump, we will suppose that no water is in the pump, and we will pump until a stream issues from the spout. The various stages are represented diagrammatically by Figure 133. In (1) the entire pump is empty of water but full of air at atmospheric pressure, and both valves are closed. In (2) the plunger is being raised and is lifting the column of air that rests on it. The air and water in the inlet pipe, being thus partially relieved of downward pressure, are pushed up by the atmospheric pressure on the surface of the water in the well. When the piston moves downward as in (3), the valve in the pipe closes by its own weight, and the air in the cylinder escapes through the valve in the plunger. In (4) the piston is again rising, repeating the process of (2). In (5) the process of (3) is being repeated, but water instead of air is escaping through the valve in the plunger. In (6) the process of (2) is being repeated, but the water has reached the spout and is flowing out.

[Illustration: FIG. 133. Diagram of the process of pumping.]

After the pump is in condition (6), motion of the plunger is followed by a more or less regular discharge of water through the spout, and the quantity of water which gushes forth depends upon the speed with which the piston is moved. A strong man giving quick strokes can produce a large flow; a child, on the other hand, is able to produce only a thin stream. Whoever pumps must exert sufficient force to lift the water from the surface of the well to the spout exit. For this reason the pump has received the name of *lifting pump*.

[Illustration: FIG. 134.--Force pump.]

184. The Force Pump. In the common pump, water cannot not be raised higher than the spout. In many cases it is desirable to force water considerably above the pump itself, as, for instance, in the fire

CHAPTER XVIII

hose; under such circumstances a type of pump is employed which has received the name of *force pump*. This differs but little from the ordinary lift pump, as a reference to Figure 134 will show. Here both valves are placed in the cylinder, and the piston is solid, but the principle is the same as in the lifting pump.

An upward motion of the plunger allows water to enter the cylinder, and the downward motion of the plunger drives water through *E*. (Is this true for the lift pump as well?) Since only the downward motion of the plunger forces water through *E*, the discharge is intermittent and is therefore not practical for commercial purposes. In order to convert this intermittent discharge into a steady stream, an air chamber is installed near the discharge tube, as in Figure 135. The water forced into the air chamber by the downward-moving piston compresses the air and increases its pressure. The pressure of the confined air reacts against the water and tends to drive it out of the chamber. Hence, even when the plunger is moving upward, water is forced through the pipe because of the pressure of the compressed air. In this way a continuous flow is secured.

[Illustration: FIG 135.--The air chamber *A* insures a continuous flow of water.]

The height to which the water can be forced in the pipe depends upon the size and construction of the pump and upon the force with which the plunger can be moved. The larger the stream desired and the greater the height to be reached, the stronger the force needed and the more powerful the construction necessary.

The force pump gets its name from the fact that the moving piston drives or forces the water through the discharge tube.

185. Irrigation and Drainage. History shows that the lifting pump has been used by man since the fourth century before Christ; for many present-day enterprises this ancient form of pump is inconvenient and impracticable, and hence it has been replaced in many cases by more modern types, such as rotary and centrifugal pumps (Fig. 136). In these forms, rapidly rotating wheels lift the water and drive it onward into a discharge pipe, from which it issues with great force. There is

CHAPTER XVIII 139

neither piston nor valve in these pumps, and the quantity of water raised and the force with which it is driven through the pipes depends solely upon the size of the wheels and the speed with which they rotate.

Irrigation, or the artificial watering of land, is of the greatest importance in those parts of the world where the land is naturally too dry for farming. In the United States, approximately two fifths of the land area is so dry as to be worthless for agricultural purposes unless artificially watered. In the West, several large irrigating systems have been built by the federal government, and at present about ten million acres of land have been converted from worthless farms into fields rich in crops. Many irrigating systems use centrifugal pumps to force water over long distances and to supply it in quantities sufficient for vast agricultural needs. In many regions, the success of a farm or ranch depends upon the irrigation furnished in dry seasons, or upon man's ability to drive water from a region of abundance to a remote region of scarcity.

[Illustration: FIG. 136.--Centrifugal pump with part of the casing] cut away to show the wheel.

[Illustration: FIG. 137.--Agriculture made possible by irrigation.]

The draining of land is also a matter of considerable importance; swamps and marshes which were at one time considered useless have been drained and then reclaimed and converted into good farming land. The surplus water is best removed by centrifugal pumps, since sand and sticks which would clog the valves of an ordinary pump are passed along without difficulty by the rotating wheel.

[Illustration: FIG. 138.--Rice for its growth needs periodical flooding, and irrigation often supplies the necessary water.]

186. Camping.--Its Pleasures and its Dangers. The allurement of a vacation camp in the heart of the woods is so great as to make many campers ignore the vital importance of securing a safe water supply. A river bank may be beautiful and teeming with diversions, but if the river is used as a source of drinking water, the results will almost always be

fatal to some. The water can be boiled, it is true, but few campers are willing to forage for the additional wood needed for this apparently unnecessary requirement; then, too, boiled water does not cool readily in summer, and hence is disagreeable for drinking purposes.

The only safe course is to abandon the river as a source of drinking water, and if a spring cannot be found, to drive a well. In many regions, especially in the neighborhood of streams, water can be found ten or fifteen feet below the surface. Water taken from such a depth has filtered through a bed of soil, and is fairly safe for any purpose. Of course the deeper the well, the safer will be the water. With the use of such a pump as will be described, campers can, without grave danger, throw dish water, etc., on the ground somewhat remote from the camp; this may not injure their drinking water because the liquids will slowly seep through the ground, and as they filter downward will lose their dangerous matter. All the water which reaches the well pipes will have filtered through the soil bed and therefore will probably be safe.

But while the careless disposal of wastes may not spoil the drinking water (in the well to be described), other laws of health demand a thoughtful disposal of wastes. The malarial mosquito and the typhoid fly flourish in unhygienic quarters, and the only way to guard against their dangers is to allow them neither food nor breeding place.

The burning of garbage, the discharge of waters into cesspools, or, in temporary camps, the discharge of wastes to distant points through the agency of a cheap sewage pipe will insure safety to campers, will lessen the trials of flies and mosquitoes, and will add but little to the expense.

187. A Cheap Well for Campers. A two-inch galvanized iron pipe with a strong, pointed end containing small perforations is driven into the ground with a sledge hammer. After it has penetrated for a few feet, another length is added and the whole is driven down, and this is repeated until water is reached. A cheap pump is then attached to the upper end of the drill pipe and serves to raise the water. During the drilling, some soil particles get into the pipe through the perforations, and these cloud the water at first; but after the pipe has once been cleaned by the upward-moving water, the supply remains clear. The

CHAPTER XVIII 141

flow from such a well is naturally small; first, because water is not abundant near the surface of the earth, and second, because cheap pumps are poorly constructed and cannot raise a large amount. But the supply will usually be sufficient for the needs of simple camp life, and many a small farm uses this form of well, not only for household purposes, but for watering the cattle in winter.

If the cheapness of such pumps were known, their use would be more general for temporary purposes. The cost of material need not exceed $5 for a 10-foot well, and the driving of the pipe could be made as much a part of the camping as the pitching of the tent itself. If the camping site is abandoned at the close of the vacation, the pump can be removed and kept over winter for use the following summer in another place. In this way the actual cost of the water supply can be reduced to scarcely more than $3, the removable pump being a permanent possession. In rocky or mountain regions the driven well is not practicable, because the driving point is blunted and broken by the rock and cannot pierce the rocky beds of land.

[Illustration: FIG. 139--A driven well.]

[Illustration: FIG. 140.--Diagram showing how supplying a city with good water lessens sickness and death. The lines *b* show the relative number of people who died of typhoid fever before the water was filtered; the lines *a* show the numbers who died after the water was filtered. The figures are the number of typhoid deaths occurring yearly out of 100,000 inhabitants.]

188. Our Summer Vacation. It has been asserted by some city health officials that many cases of typhoid fever in cities can be traced to the unsanitary conditions existing in summer resorts. The drinking water of most cities is now under strict supervision, while that of isolated farms, of small seaside resorts, and of scattered mountain hotels is left to the care of individual proprietors, and in only too many instances receives no attention whatever. The sewage disposal is often inadequate and badly planned, and the water becomes dangerously contaminated. A strong, healthy person, with plenty of outdoor exercise and with hygienic habits, may be able to resist the disease germs present in the poor water supply; more often the summer guests carry back with

CHAPTER XVIII   142

them to their winter homes the germs of disease, and these gain the upper hand under the altered conditions of city and business life. It is not too much to say that every man and woman should know the source of his summer table water and the method of sewage disposal. If the conditions are unsanitary, they cannot be remedied at once, but another resort can be found and personal danger can be avoided. Public sentiment and the loss of trade will go far in furthering an effort toward better sanitation.

In the driven well, water cannot reach the spout unless it has first filtered through the soil to the depth of the driven pipe; after such a journey it is fairly safe, unless very large quantities of sewage are present; generally speaking, such a depth of soil is able to filter satisfactorily the drainage of the limited number of people which a driven well suffices to supply.

[Illustration: FIG. 141.--A deep well with the piston in the water.]

Abundant water is rarely reached at less than 75 feet, and it would usually be impossible to drive a pipe to such a depth. When a large quantity of water is desired, strong machines drill into the ground and excavate an opening into which a wide pipe can be lowered. I recently spent a summer in the Pocono Mountains and saw such a well completed. The machine drilled to a depth of 250 feet before much water was reached and to over 300 feet before a flow was obtained sufficient to satisfy the owner. The water thus obtained was to be the sole water supply of a hotel accommodating 150 persons; the proprietor calculated that the requirements of his guests, for bath, toilet, laundry, kitchen, etc., and the domestics employed to serve them, together with the livery at their disposal, demanded a flow of 10 gallons per minute. The ground was full of rock and difficult to penetrate, and it required 6 weeks of constant work for two skilled men to drill the opening, lower the suction pipe, and install the pump, the cost being approximately $700.

[Illustration: FIG. 142.--Showing how drinking water can be contaminated from cesspool *(c)* and wash water *(w)*.]

The water from such a well is safe and pure except under the conditions represented in Figure 142. If sewage or slops be poured upon the ground in the neighborhood of the well, the liquid will seep through the ground and some may make its way into the pump before it has been purified by the earth. The impure liquid will thus contaminate the otherwise pure water and will render it decidedly harmful. For absolute safety the sewage discharge should be at least 75 feet from the well, and in large hotels, where there is necessarily a large quantity of sewage, the distance should be much greater. As the sewage seeps through the ground it loses its impurities, but the quantity of earth required to purify it depends upon its abundance; a small depth of soil cannot take care of an indefinite amount of sewage. Hence, the greater the number of people in a hotel, or the more abundant the sewage, the greater should be the distance between well and sewer.

By far the best way to avoid contamination is to see to it that the sewage discharges into the ground *below* the well; that is, to dig the well in such a location that the sewage drainage will be away from the well.

In cities and towns and large summer communities, the sewage of individual buildings drains into common tanks erected at public expense; the contents of these are discharged in turn into harbors and streams, or are otherwise disposed of at great expense, although they contain valuable substances. It has been estimated that the drainage or sewage of England alone would be worth $ 80,000,000 a year if used as fertilizer.

A few cities, such as Columbus and Cleveland, Ohio, realize the need of utilizing this source of wealth, and by chemical means deodorize their sewage and change it into substances useful for agricultural and industrial purposes. There is still a great deal to be learned on this subject, and it is possible that chemically treated sewage may be made a source of income to a community rather than an expense.

189. Pumps which Compress Air. The pumps considered in the preceding Sections have their widest application in agricultural districts, where by means of them water is raised to the surface of the

CHAPTER XVIII

earth or is pumped into elevated tanks. From a commercial and industrial standpoint a most important class of pump is that known as the compression type; in these, air or any other gas is compressed rather than rarefied.

Air brakes and self-opening and self-closing doors on cars are operated by means of compression pumps. The laying of bridge and pier foundations, in fact all work which must be done under water, is possible only through the agency of compression pumps. Those who have visited mines, and have gone into the heart of the underground labyrinth, know how difficult it is for fresh air to make its way to the miners. Compression pumps have eliminated this difficulty, and to-day fresh air is constantly pumped into the mines to supply the laborers there. Agricultural methods also have been modified by the compression pump. The spraying of trees (Fig. 143), formerly done slowly and laboriously, is now a relatively simple matter.

[Illustration: FIG. 143.--Spraying trees by means of a compression pump.]

190. The Bicycle Pump. The bicycle pump is the best known of all compression pumps. Here, as in other pumps of its type, the valves open inward rather than outward. When the piston is lowered, compressed air is driven through the rubber tubing, pushes open an inward-opening valve in the tire, and thus enters the tire. When the piston is raised, the lower valve closes, the upper valve is opened by atmospheric pressure, and air from outside enters the cylinder; the next stroke of the piston drives a fresh supply of air into the tire, which thus in time becomes inflated. In most cheap bicycle pumps, the piston valve is replaced by a soft piece of leather so attached to the piston that it allows air to slip around it and into the cylinder, but prevents its escape from the cylinder (Fig. 144).

[Illustration: FIG. 144.--The bicycle foot pump.]

191. How a Man works under Water. Place one end of a piece of glass tube in a vessel of water and notice that the water rises in the tube (Fig. 145). Blow into the tube and see whether you can force the water wholly or partially down the tube. If the tube is connected to a small

compression pump, sufficient air can be sent into the tube to cause the water to sink and to keep the tube permanently clear of water. This is, in brief, the principle employed for work under water. A compression pump forces air through a tube into the chamber in which men are to work (Fig. 146). The air thus furnished from above supplies the workmen with oxygen, and by its pressure prevents water from entering the chamber. When the task has been completed, the chamber is raised and later lowered to a new position.

[Illustration: FIG. 145.--Water does not enter the tube as long as we blow into it.]

Figure 147 shows men at work on a bridge foundation. Workmen, tools, and supplies are lowered in baskets through a central tube *BC* provided with an air chamber *L*, having air-tight gates at *A* and *A'*. The gate *A* is opened and workmen enter the air chamber. The gate *A* is then closed and the gate *A'* is opened slowly to give the men time to get accustomed to the high pressure in *B*, and then the men are lowered to the bottom. Excavated earth is removed in a similar manner. Air is supplied through a tube *DD*. Such an arrangement for work under water is called a caisson. It is held in position by a mass of concrete *EE*.

[Illustration: FIG. 146--The principle of work under water.]

[Illustration: FIG. 147--Showing how men can work under water.]

In many cases men work in diving suits rather than in caissons; these suits are made of rubber except for the head piece, which is of metal provided with transparent eyepieces. Air is supplied through a flexible tube by a compression pump. The diver sometimes carries on his back a tank of compressed air, from which the air escapes through a tube to the space between the body and the suit. When the air has become foul, the diver opens a valve in his suit and allows it to pass into the water, at the same time admitting a fresh supply from the tank. The valve opens outward from the body, and hence will allow of the exit of air but not of the entrance of water. When the diver ceases work and desires to rise to the surface, he signals and is drawn up by a rope attached to the suit.

192. Combination of Pumps. In many cases the combined use of both exhaust and compression pumps is necessary to secure the desired result; as, for example, in pneumatic dispatch tubes. These are employed in the transportation of letters and small packages from building to building or between parts of the same building. A pump removes air from the part of the tube ahead of the package, and thus reduces the resistance, while a compression pump forces air into the tube behind the package and thus drives it forward with great speed.

CHAPTER XIX

THE WATER PROBLEM OF A LARGE CITY

193. It is by no means unusual for the residents of a large city or town to receive through the newspapers a notification that the city water supply is running low and that economy should be exercised in its use. The problem of supplying a large city with an abundance of pure water is among the most difficult tasks which city officials have to perform, and is one little understood and appreciated by the average citizen.

Intense interest in personal and domestic affairs is natural, but every citizen, rich or poor, should have an interest in civic affairs as well, and there is no better or more important place to begin than with the water supply. One of the most stirring questions in New York to-day has to do with the construction of huge aqueducts designed to convey to the residents of the city, water from the distant Catskill Mountains. The growth of the population has been so phenomenally rapid that the combined output of all available near-by sources does not suffice to meet the increasing consumption.

Where does your city obtain its water? Does it bring it to its reservoirs in the most economic way possible, and is there any legitimate excuse for the scarcity of water which many communities face in dry seasons?

194. Two Possibilities. Sometimes a city is fortunate enough to be situated near hills and mountains through which streams flow, and in that case the water problem is simple. In such a case all that is

CHAPTER XIX   147

necessary is to run pipes, usually underground, from the elevated lakes or streams to the individual houses, or to common reservoirs from which it is distributed to the various buildings.

[Illustration: FIG. 148.--The elevated mountain lake serves as a source of water.]

Figure 148 illustrates in a simple way the manner in which a mountain lake may serve to supply the inhabitants of a valley. The city of Denver, for example, is surrounded by mountains abounding in streams of pure, clear water; pipes convey the water from these heights to the city, and thus a cheap and adequate flow is obtained. Such a system is known as the gravity system. The nearer and steeper the elevation, the greater the force with which the water flows through the valley pipes, and hence the stronger the discharge from the faucets.

Relatively few cities and towns are so favorably situated as regards water; more often the mountains are too distant, or the elevation is too slight, to be of practical value. Cities situated in plains and remote from mountains are obliged to utilize the water of such streams as flow through the land, forcing it to the necessary height by means of pumps. Streams which flow through populated regions are apt to be contaminated, and hence water from them requires public filtration. Cities using such a water supply thus have the double expense of pumping and filtration.

195. The Pressure of Water. No practical business man would erect a turbine or paddle wheel without calculating in advance the value of his water power. The paddle wheel might be so heavy that the stream could not turn it, or so frail in comparison with the water force that the stream would destroy it. In just as careful a manner, the size and the strength of municipal reservoirs and pumps must be calculated. The greater the quantity of water to be held in the reservoir, the heavier are the walls required; the greater the elevation of the houses, the stronger must be the pumps and the engines which run them.

In order to understand how these calculations are made, we must study the physical characteristics of water just as we studied the

CHAPTER XIX

physical characteristics of air.

When we measure water, we find that 1 cubic foot of it weighs about 62.5 pounds; this is equivalent to saying that water 1 foot deep presses on the bottom of the containing vessel with a force of 62.5 pounds to the square foot. If the water is 2 feet deep, the load supported by the vessel is doubled, and the pressure on each square foot of the bottom of the vessel will be 125 pounds, and if the water is 10 feet deep, the load borne by each square foot will be 625 pounds. The deeper the water, the greater will be the weight sustained by the confining vessel and the greater the pressure exerted by the water.

[Illustration: FIG. 149.--Water 1 foot deep exerts a pressure of 62.5 pounds a square foot.]

Since the pressure borne by 1 square foot of surface is 62.5 pounds, the pressure supported by 1 square inch of surface is 1/144 of 62.5 pounds, or .43 pound, nearly 1/2 pound. Suppose a vessel held water to the depth of 10 feet, then upon every square inch of the bottom of that vessel there would be a pressure of 4.34 pounds. If a one-inch tap were inserted in the bottom of the vessel so that the water flowed out, it would gush forth with a force of 4.34 pounds. If the water were 20 feet deep, the force of the outflowing water would be twice as strong, because the pressure would be doubled. But the flow would not remain constant, because as the water leaves the outlet, less and less of it remains in the vessel, and hence the pressure gradually sinks and the flow drops correspondingly.

In seasons of prolonged drought, the streams which feed a city reservoir are apt to contain less than the usual amount of water, hence the level of the water supply sinks, the pressure at the outlet falls, and the force of the outflowing water is lessened (Fig. 150).

[Illustration: FIG. 150.--The pressure at an outlet decreases as the level of the water supply sinks.]

196. Why the Water Supply is not uniform in All Parts of the City. In the preceding Section, we saw that the flow from a faucet depends upon the height of the reserve water above the tap. Houses on a level with

the main supply pipes (Figs. 148 and 151) have a strong flow because the water is under the pressure of a column A; houses situated on elevation B have less flow, because the water is under the pressure of a shorter column B; and houses at a considerable elevation C have a less rapid flow corresponding to the diminished depth (C).

Not only does the flow vary with the elevation of the house, but it varies with the location of the faucet within the house. Unless the reservoir is very high, or the pumps very powerful, the flow on the upper floors is noticeably less than that in the cellar, and in the upper stories of some high building the flow is scarcely more than a feeble trickle.

[Illustration: FIG. 151.--Water pressure varies in different parts of a water system.]

When the respective flows at A, B, and C (Fig. 151) are measured, they are found to be far lower than the pressures which columns of water of the heights A, B, and C have been shown by actual demonstration to exert. This is because water, in flowing from place to place, expends force in overcoming the friction of the pipes and the resistance of the air. The greater the distance traversed by the water in its journey from reservoir to faucet, the greater the waste force and the less the final flow.

In practice, large mains lead from the reservoir to the city, smaller mains convey the water to the various sections of the city, and service pipes lead to the individual house taps. During this long journey, considerable force is expended against friction, and hence the flow at a distance from the reservoir falls to but a fraction of its original strength. For this reason, buildings situated near the main supply have a much stronger flow (Fig. 152) than those on the same level but remote from the supply. Artificial reservoirs are usually constructed on the near outskirts of a town in order that the frictional force lost in transmission may be reduced to a minimum.

[Illustration: FIG. 152.--The more distant the fountain, the weaker the flow.]

CHAPTER XIX

150

In the case of a natural reservoir, such as an elevated lake or stream, the distance cannot be planned or controlled. New York, for example, will secure an abundance of pure water from the Catskill Mountains, but it will lose force in transmission. Los Angeles is undertaking one of the greatest municipal projects of the day. Huge aqueducts are being built which will convey pure mountain water a distance of 250 miles, and in quantities sufficient to supply two million people. According to calculations, the force of the water will be so great that pumps will not be needed.

197. Why Water does not always flow from a Faucet. Most of us have at times been annoyed by the inability to secure water on an upper story, because of the drawing off of a supply on a lower floor. During the working hours of the day, immense quantities of water are drawn off from innumerable faucets, and hence the quantity in the pipes decreases considerably unless the supply station is able to drive water through the vast network of pipes as fast as it is drawn off. Buildings at a distance from the reservoir suffer under such circumstances, because while the diminished pressure is ordinarily powerful enough to supply the lower floors, it is frequently too weak to force a continuous stream to high levels. At night, however, and out of working hours, few faucets are open, less water is drawn off at any one time, and the intricate pipes are constantly full of water under high pressure. At such times, a good flow is obtainable even on the uppermost floors.

In order to overcome the disadvantage of a decrease in flow during the day, standpipes (Fig. 153) are sometimes placed in various sections. These are practically small steel reservoirs full of water and connecting with the city pipes. During "rush" hours, water passes from these into the communicating pipes and increases the available supply, while during the night, when the faucets are turned off, water accumulates in the standpipe against the next emergency (Figs. 151 and 154). The service rendered by the standpipe is similar to that of the air cushion discussed in Section 184.

[Illustration: FIG. 153.--A standpipe.]

198. The Cost of Water. In the gravity system, where an elevated lake or stream serves as a natural reservoir, the cost of the city's

CHAPTER XIX

waterworks is practically limited to the laying of pipes. But when the source of the supply is more or less on a level with the surrounding land, the cost is great, because the supply for the entire city must either be pumped into an artificial reservoir, from which it can be distributed, or else must be driven directly through the mains (Fig. 154).

[Illustration: FIG. 154.--Water must be got to the houses by means of pumps.]

A gallon of water weighs approximately 8.3 pounds, and hence the work done by a pump in raising a gallon of water to the top of an average house, an elevation of 50 feet, is 8.3×50, or 415 foot pounds. A small manufacturing town uses at least 1,000,000 gallons daily, and the work done by a pump in raising that amount to an elevation of 50 feet would be $8.3 \times 1,000,000 \times 50$, or 415,000,000 foot pounds.

The total work done during the day by the pump, or the engine driving the pump, is 415,000,000 foot pounds, and hence the work done during one hour would be 1/24 of 415,000,000, or 17,291,666 foot pounds; the work done in one minute would be 1/60 of 17,291,666, or 288,194 foot pounds, and the work done each second would be 1/60 of 288,194, or 4803 foot pounds.

A 1-H.P. engine does 550 foot pounds of work each second, and therefore if the pump is to be operated by an engine, the strength of the latter would have to be 8.7 H.P. An 8.7-H.P. pumping engine working at full speed every second of the day and night would be able to supply the town with the necessary amount of water. When, however, we consider the actual height to which the water is raised above the pumping station, and the extra pumping which must be done in order to balance the frictional loss, it is easy to understand that in actual practice a much more powerful engine would be needed. The larger the piston and the faster it works, the greater is the quantity of water raised at each stroke, and the stronger must be the engine which operates the pump.

In many large cities there is no one single pumping station from which supplies run to all parts of the city, but several pumping stations are

CHAPTER XIX 152

scattered throughout the city, and each of them supplies a restricted territory.

199. The Bursting of Dams and Reservoirs. The construction of a safe reservoir is one of the most important problems of engineers. In October, 1911, a town in Pennsylvania was virtually wiped out of existence because of the bursting of a dam whose structure was of insufficient strength to resist the strain of the vast quantity of water held by it. A similar breakage was the cause of the fatal Johnstown flood in 1889, which destroyed no less than seven towns, and in which approximately 2000 persons are said to have lost their lives.

Water presses not only on the bottom of a vessel, but upon the sides as well; a bucket leaks whether the hole is in its side or its bottom, showing that water presses not only downward but outward. Usually a leak in a dam or reservoir occurs near the bottom. Weak spots at the top are rare and easily repaired, but a leak near the bottom is usually fatal, and in the case of a large reservoir the outflowing water carries death and destruction to everything in its path.

If the leak is near the surface, as at *a* (Fig. 155), the water issues as a feeble stream, because the pressure against the sides at that level is due solely to the relatively small height of water above *a* (Section 195). If the leak is lower, as at *b*, the issuing stream is stronger and swifter, because at that level the outward pressure is much greater than at *a*, the increase being due to the fact that the height of the water above *b* is greater than that above *a*. If the leak is quite low, as at *c*, the issuing stream has a still greater speed and strength, and gushes forth with a force determined by the height of the water above *c*.

[Illustration: FIG. 155.--The flow from an opening depends upon the height of water above the opening.]

The dam at Johnstown was nearly 1/2 mile wide, and 40 feet high, and so great was the force and speed of the escaping stream that within an hour after the break had occurred, the water had traveled a distance of 18 miles, and had destroyed property to the value of millions of dollars.

CHAPTER XIX

If a reservoir has a depth of 100 feet, the pressure exerted upon each square foot of its floor is 62.5 × 100, or 6250 pounds; the weight therefore to be sustained by every square foot of the reservoir floor is somewhat more than 3 tons, and hence strong foundations are essential. The outward lateral pressure at a depth of 25 feet would be only one fourth as great as that on the bottom--hence the strain on the sides at that depth would be relatively slight, and a less powerful construction would suffice. But at a depth of 50 feet the pressure on the sides would be one half that of the floor pressure, or 1-1/2 tons. At a depth of 75 feet, the pressure on the sides would be three quarters that on the bottom, or 2-1/4 tons. As the bottom of the reservoir is approached, the pressure against the sides increases, and more powerful construction becomes necessary.

Small elevated tanks, like those of the windmill, frequently have heavy iron bands around their lower portion as a protection against the extra strain.

Before erecting a dam or reservoir, the maximum pressure to be exerted upon every square inch of surface should be accurately calculated, and the structure should then be built in such a way that the varying pressure of the water can be sustained. It is not sufficient that the bottom be strong; the sides likewise must support their strain, and hence must be increased in strength with depth. This strengthening of the walls is seen clearly in the reservoir shown in Figure 152. The bursting of dams and reservoirs has occasioned the loss of so many lives, and the destruction of so much property, that some states are considering the advisability of federal inspection of all such structures.

[Illustration: FIG. 156.--The lock gates must be strong in order to withstand the great pressure of the water against them.]

200. The Relation of Forests to the Water Supply. When heavy rains fall on a bare slope, or when snow melts on a barren hillside, a small amount of the water sinks into the ground, but by far the greater part of it runs off quickly and swells brooks and streams, thus causing floods and freshets.

When, however, rain falls on a wooded slope, the action is reversed; a small portion runs off, while the greater portion sinks into the soft earth. This is due partly to the fact that the roots of trees by their constant growth keep the soil loose and open, and form channels, as it were, along which the water can easily run. It is due also to the presence on the ground of decaying leaves and twigs, or humus. The decaying vegetable matter which covers the forest floor acts more or less as a sponge, and quickly absorbs falling rain and melting snow. The water which thus passes into the humus and the soil beneath does not remain there, but slowly seeps downward, and finally after weeks and months emerges at a lower level as a stream. Brooks and springs formed in this way are constant feeders of rivers and lakes.

In regions where the land has been deforested, the rivers run low in season of prolonged drought, because the water which should have slowly seeped through the soil, and then supplied the rivers for weeks and months, ran off from the barren slopes in a few days.

Forests not only lessen the danger of floods, but they conserve our waterways, preventing a dangerous high-water mark in the season of heavy rains and melting snows, and then preventing a shrinkage in dry seasons when the only feeders of the rivers are the underground sources. In the summer of 1911, prolonged drought in North Carolina lowered the rivers to such an extent that towns dependent upon them suffered greatly. The city of Charlotte was reduced for a time to a practically empty reservoir; washing and bathing were eliminated, machinery dependent upon water-power and steam stood idle, and every glass of water drunk was carefully reckoned. Thousands of gallons of water were brought in tanks from neighboring cities, and were emptied into the empty reservoir from whence it trickled slowly through the city mains. The lack of water caused not only personal inconvenience and business paralysis, but it occasioned real danger of disease through unflushed sewers and insufficiently drained pipes.

The conservation of the forest means the conservation of our waterways, whether these be used for transportation or as sources of drinking water.

CHAPTER XX

MAN'S CONQUEST OF SUBSTANCES

201. Chemistry. Man's mechanical inventions have been equaled by his chemical researches and discoveries, and by the application he has made of his new knowledge.

The plain cotton frock of our grandmothers had its death knell sounded a few years ago, when John Mercer showed that cotton fabrics soaked in caustic soda assumed under certain conditions a silky sheen, and when dyed took on beautiful and varied hues. The demonstration of this simple fact laid the foundation for the manufacture of a vast variety of attractive dress materials known as mercerized cotton.

Possibly no industry has been more affected by chemical discovery than that of dyeing. Those of us who have seen the old masterpieces in painting, or reproductions of them, know the softness, the mellowness, the richness of tints employed by the old masters. But if we look for the brilliancy and variety of color seen in our own day, the search will be fruitless, because these were unknown until a half century ago. Up to that time, dyes were few in number and were extracted solely from plants, principally from the indigo and madder plants. But about the year 1856 it was discovered that dyes in much greater variety and in purer form could be obtained from coal tar. This chemical production of dyes has now largely supplanted the original method, and the industry has grown so rapidly that a single firm produced in one year from coal tar a quantity of indigo dye which under the natural process of plant extraction would have required a quarter million acres of indigo plant.

The abundance and cheapness of newspapers, coarse wrapping papers, etc., is due to the fact that man has learned to substitute wood for rags in the manufacture of paper. Investigation brought out the fact that wood contained the substance which made rags valuable for paper making. Since the supply of rags was far less than the demand, the problem of the extraction from wood of the paper-forming substance was a vital one. From repeated trials, it was found that caustic soda when heated with wood chips destroyed everything in the

CHAPTER XX

wood except the desired substance, cellulose; this could be removed, bleached, dried, and pressed into paper. The substitution of wood for rags has made possible the daily issue of newspapers, for the making of which sufficient material would not otherwise have been available. When we reflect that a daily paper of wide circulation consumes ten acres of wood lot per day, we see that all the rags in the world would be inadequate to meet this demand alone, to say nothing of periodicals, books, tissue paper, etc.

Chemistry plays a part in every phase of life; in the arts, the industries, the household, and in the body itself, where digestion, excretion, etc., result from the action of the bodily fluids upon food. The chemical substances of most interest to us are those which affect us personally rather than industrially; for example, soap, which cleanses our bodies, our clothing, our household possessions; washing soda, which lightens laundry work; lye, which clears out the drain pipe clogged with grease; benzine, which removes stains from clothing; turpentine, which rids us of paint spots left by careless workmen; and hydrogen peroxide, which disinfects wounds and sores.

In order to understand the action of several of these substances we must study the properties of two groups of chemicals--known respectively as acids and bases; the first of these may be represented by vinegar, sulphuric acid, and oxalic acid; and the second, by ammonia, lye, and limewater.

202. Acids. All of us know that vinegar and lemon juice have a sour taste, and it is easy to show that most acids are characterized by a sour taste. If a clean glass rod is dipped into very dilute acid, such as acetic, sulphuric, or nitric acid, and then lightly touched to the tongue, it will taste sour. But the best test of an acid is by sight rather than by taste, because it has been found that an acid is able to discolor a plant substance called litmus. If paper is soaked in a litmus solution until it acquires the characteristic blue hue of the plant substance, and is then dried thoroughly, it can be used to detect acids, because if it comes in contact with even the minutest trace of acid, it loses its blue color and assumes a red tint. Hence, in order to detect the presence of acid in a substance, one has merely to put some of the substance on blue litmus paper, and note whether or not the latter changes color. This

CHAPTER XX 157

test shows that many of our common foods contain some acid; for example, fruit, buttermilk, sour bread, and vinegar.

The damage which can be done by strong acids is well known; if a jar of sulphuric acid is overturned, and some of it falls on the skin, it eats its way into the flesh and leaves an ugly sore; if it falls on carpet or coat, it eats its way into the material and leaves an unsightly hole. The evil results of an accident with acid can be lessened if we know just what to do and do it quickly, but for this we must have a knowledge of bases, the second group of chemicals.

203. Bases. Substances belonging to this group usually have a bitter taste and a slimy, soapy feeling. For our present purposes, the most important characteristic of a base is that it will neutralize an acid and in some measure hinder the damage effected by the former. If, as soon as an acid has been spilled on cloth, a base, such as ammonia, is applied to the affected region, but little harm will be done. In your laboratory experiments you may be unfortunate enough to spill acid on your body or clothing; if so, quickly apply ammonia. If you delay, the acid does its work, and there is no remedy. If soda (a base) touches black material, it discolors it and leaves an ugly brown spot; but the application of a little acid, such as vinegar or lemon juice, will often restore the original color and counteract the bad effects of the base. Limewater prescribed by physicians in cases of illness is a well-known base. This liquid neutralizes the too abundant acids present in a weak system and so quiets and tones the stomach.

The interaction of acids and bases may be observed in another way. If blue litmus paper is put into an acid solution, its color changes to red; if now the red litmus paper is dipped into a base solution, caustic soda, for example, its original color is partially restored. What the acid does, the base undoes, either wholly or in part. Bases always turn red litmus paper blue.

Bases, like acids, are good or bad according to their use; if they come in contact with cloth, they eat or discolor it, unless neutralized by an acid. But this property of bases, harmful in one way, is put to advantage in the home, where grease is removed from drainpipe and sink by the application of lye, a strong base. If the lye is too concentrated, it will not only eat the grease, but will corrode the metal piping; it is easy, however, to dilute base solutions to such a degree that they will not affect piping, but will remove grease. Dilute ammonia is used in almost every home and is an indispensable domestic servant; diluted sufficiently, it is invaluable in the washing of delicate fabrics and in the removing of stains, and in a more concentrated form it is helpful as a smelling salt in cases of fainting.

Some concentrated bases are so powerful in their action on grease, cloth, and metal that they have received the designation *caustic*, and are ordinarily known as caustic soda, caustic potash (lye), and caustic lime. These more active bases are generally called alkalies in distinction from the less active ones.

204. Neutral Substances. To any acid solution add gradually a small quantity of a base, and test the mixture from time to time with blue litmus paper; at first the paper will turn red quickly, but as more and more of the base is added to the solution, it has less and less effect on the blue litmus paper, and finally a point is reached when a fresh strip of blue paper will not be affected. Such a result indicates infallibly the absence of any acid qualities in the solution. If now red litmus paper is tested in the same solution, its color also will remain unchanged; such a result indicates infallibly the absence of any basic quality. The solution has the characteristic property of neither acid nor base and is said to be neutral.

If to the neutral solution an extra portion of base is added, so that there is an excess of base over acid, the neutralization is overbalanced and the red paper turns blue. If to the neutral solution an extra portion of acid is added, so that there is an excess of acid over base, the neutralization is overbalanced in the opposite direction, and the solution acquires acid characteristics.

Most acids and bases will eat and corrode and discolor, while neutral substances will not; it is for this reason that soap, a slightly alkaline substance, is the safest cleansing agent for laundry, bath, and general work. Good soaps, being carefully made, are so nearly neutral that they will not fade the color out of clothing; the cheap soaps are less carefully prepared and are apt to have a strong excess of the base

CHAPTER XX 159

ingredient; such soaps are not safe for delicate work.

205. Soap. If we gather together scrapings of lard, butter, bits of tallow from burned-out candles, scraps of waste fat, or any other sort of grease, and pour a strong solution of lye over the mass, a soft soapy substance is formed. In colonial times, every family made its own supply of soap, utilizing, for that purpose, household scraps often regarded by the housekeeper of to-day as worthless. Grease and fat were boiled with water and hardwood ashes, which are rich in lye, and from the mixture came the soft soap used by our ancestors. In practice, the wood ashes were boiled in water, which was then strained off, and the resulting filtrate, or lye, was mixed with the fats for soap making.

Most fats contain a substance of an acid nature, and are decomposed by the action of bases such as caustic soda and caustic potash. The acid component of the grease partially neutralizes the base, and a new substance is formed, namely, soap.

With the advance of civilization the labor of soap making passed from the home to the factory, very much as bread making has done in our own day. Different varieties of soaps appeared, of which the hard soap was the most popular, owing to the ease with which it could be transported. Within the last few years liquid soaps have come into favor, especially in schools, railroad stations, and other public places, where a cake of soap would be handled by many persons. By means of a simple device (Fig. 157), the soap escapes from a receptacle when needed. The mass of the soap does not come in contact with the skin, and hence the spread of contagious skin diseases is lessened.

[Illustration: FIG. 157.--Liquid soap container.]

Commercial soaps are made from a great variety of substances, such as tallow, lard, castor oil, coconut oil, olive oil, etc.; or in cheaper soaps, from rosin, cottonseed oil, and waste grease. The fats which go to waste in our garbage could be made a source of income, not only to the housewife, but to the city. In Columbus, Ohio, garbage is used as a source of revenue; the grease from the garbage being sold for soap making, and the tankage (Section 188) for fertilizer.

CHAPTER XX

160

206. Why Soap Cleans. The natural oil of the skin catches and retains dust and dirt, and makes a greasy film over the body. This cannot be removed by water alone, but if soap is used and a generous lather is applied to the skin, the dirt is "cut" and passes from the body into the water. Soap affects a grease film and water very much as the white of an egg affects oil and water. These two liquids alone do not mix, the oil remaining separate on the surface of the water; but if a small quantity of white of egg is added, an emulsion is formed, the oil separating into minute droplets which spread through the water. In the same way, soap acts on a grease film, separating it into minute droplets which leave the skin and spread through the water, carrying with them the dust and dirt particles. The warmer the water, the better will be the emulsion, and hence the more effective the removal of dirt and grease. This explanation holds true for the removal of grease from any surface, whether of the body, clothing, furniture, or dishes.

207. Washing Powders. Sometimes soap refuses to form a lather and instead cakes and floats as a scum on the top of the water; this is not the fault of the soap but of the water. As water seeps through the soil or flows over the land, it absorbs and retains various soil constituents which modify its character and, in some cases, render it almost useless for household purposes. Most of us are familiar with the rain barrel of the country house, and know that the housewife prefers rain water for laundry and general work. Rain water, coming as it does from the clouds, is free from the chemicals gathered by ground water, and is hence practically pure. While foreign substances do not necessarily injure water for drinking purposes (Section 69), they are often of such a nature as to prevent soap from forming an emulsion, and hence from doing its work. Under such circumstances the water is said to be hard, and soap used with it is wasted. Even if water is only moderately hard, much soap is lost. The substances which make water hard are calcium and magnesium salts. When soap is put into water containing one or both of these, it combines with the salts to form sticky insoluble scum. It is therefore not free to form an emulsion and to remove grease. As a cleansing agent it is valueless. The average city supply contains so little hardness that it is satisfactory for toilet purposes; but in the laundry, where there is need for the full effect of the soap, and where the slightest loss would aggregate a great deal in the course of time, something must be done to counteract the hardness. The addition of

CHAPTER XX

soda, or sodium carbonate to the water will usually produce the desired effect. Washing soda combines with calcium and magnesium and prevents them from uniting with soap. The soap is thus free to form an emulsion, just as in ordinary water. Washing powders are sometimes used instead of washing soda. Most washing powders contain, in addition to a softening agent, some alkali, and hence a double good is obtained from their use; they not only soften the water and allow the soap to form an emulsion, but they also, through their alkali content, cut the grease and themselves act as cleansers. In some cities where the water is very hard, as in Columbus, Ohio, it is softened and filtered at public expense, before it leaves the reservoirs. But even under these circumstances, a moderate use of washing powder is general in laundry work.

If washing powder is put on clothes dry, or is thrown into a crowded tub, it will eat the clothes before it has a chance to dissolve in the water. The only safe method is to dissolve the powder before the clothes are put into the tub. The trouble with our public laundries is that many of them are careless about this very fact, and do not take time to dissolve the powder before mixing it with the clothes.

The strongest washing powder is soda, and this cheap form is as good as any of the more expensive preparations sold under fancy names. Borax is a milder powder and is desirable for finer work.

One of the most disagreeable consequences of the use of hard water for bathing is the unavoidable scum which forms on the sides of bathtub and washbowl. The removal of the caked grease is difficult, and if soap alone is used, the cleaning of the tub requires both patience and hard scrubbing. The labor can be greatly lessened by moistening the scrubbing cloth with turpentine and applying it to the greasy film, which immediately dissolves and thus can be easily removed. The presence of the scum can be largely avoided by adding a small amount of liquid ammonia to the bath water. But many persons object to this; hence it is well to have some other easy method of removing the objectionable matter.

208. To remove Stains from Cloth. While soap is, generally speaking, the best cleansing agent, there are occasions when other substances

CHAPTER XX 162

can be used to better advantage. For example, grease spots on carpet and non-washable dress goods are best removed by the application of gasoline or benzine. These substances dissolve the grease, but do not remove it from the clothing; for that purpose a woolen cloth should be laid under the stain in readiness to absorb the benzine and the grease dissolved in it. If the grease is not absorbed while in solution, it remains in the clothing and after the evaporation of the benzine reappears in full force.

Cleaners frequently clean suits by laying a blotter over a grease spot and applying a hot iron; the grease, when melted by the heat, takes the easiest way of spreading itself and passes from cloth to blotter.

209. Salts. A neutral liquid formed as in Section 204, by the action of hydrochloric acid and the alkali solution of caustic soda, has a brackish, salty taste, and is, in fact, a solution of salt. This can be demonstrated by evaporating the neutral liquid to dryness and examining the residue of solid matter, which proves to be common salt.

When an acid is mixed with a base, the result is a substance more or less similar in its properties to common salt; for this reason all compounds formed by the neutralization of an acid with a base are called salts. If, instead of hydrochloric acid (HCl), we use an acid solution of potassium tartrate, and if instead of caustic soda we use bicarbonate of soda (baking soda), the result is a brackish liquid as before, but the salt in the liquid is not common salt, but Rochelle salt. Different combinations of acids and bases produce different salts. Of all the vast group of salts, the most abundant as well as the most important is common salt, known technically as sodium chloride because of its two constituents, sodium and chlorine.

We are not dependent upon neutralization for the enormous quantities of salt used in the home and in commerce. It is from the active, restless seas of the present, and from the dead seas of the prehistoric past that our vast stores of salt come. The waters of the Mediterranean and of our own Great Salt Lake are led into shallow basins, where, after evaporation by the heat of the sun, they leave a residue of salt. By far the largest quantity of salt, however, comes from the seas which

CHAPTER XX
163

no longer exist, but which in far remote ages dried up and left behind them their burden of salt. Deposits of salt formed in this way are found scattered throughout the world, and in our own country are found in greatest abundance in New York. The largest salt deposit known has a depth of one mile and exists in Germany.

Salt is indispensable on our table and in our kitchen, but the amount of salt used in this way is far too small to account for a yearly consumption of 4,000,000 tons in the United States alone. The manufacture of soap, glass, bleaching powders, baking powders, washing soda, and other chemicals depends on salt, and it is for these that the salt beds are mined.

210. Baking Soda. Salt is by all odds the most important sodium compound. Next to it come the so-called carbonates: first, sodium carbonate, which is already familiar to us as washing soda; and second, sodium bicarbonate, which is an ingredient of baking powders. These are both obtained from sodium chloride by relatively simple means; that is, by treating salt with the base, ammonia, and with carbon dioxide.

Washing soda has already been discussed. Since baking powders in some form are used in almost all homes for the raising of cake and pastry dough, it is essential that their helpful and harmful qualities be clearly understood.

The raising of dough by means of baking soda--bicarbonate of soda--is a very simple process. When soda is heated, it gives off carbon dioxide gas; you can easily prove this for yourself by burning a little soda in a test tube, and testing the escaping gas in a test tube of limewater. When flour and water alone are kneaded and baked in loaves, the result is a mass so compact and hard that human teeth are almost powerless to crush and chew it. The problem is to separate the mass of dough or, in other words, to cause it to rise and lighten. This can be done by mixing a little soda in the flour, because the heat of the oven causes the soda to give off bubbles of gas, and these in expanding make the heavy mass slightly porous. Bread is never lightened with soda because the amount of gas thus given off is too small to convert heavy compact bread dough into a spongy mass; but biscuit and cake, being by nature less compact and heavy, are sufficiently lightened by the gas given off from soda.

But there is one great objection to the use of soda alone as a leavening agent. After baking soda has lost its carbon dioxide gas, it is no longer baking soda, but is transformed into its relative, washing soda, which has a disagreeable taste and is by no means desirable for the stomach.

Man's knowledge of chemicals and their effect on each other has enabled him to overcome this difficulty and, at the same time, to retain the leavening effect of the baking soda.

211. Baking Powders. If some cooking soda is put into lemon juice or vinegar, or any acid, bubbles of gas immediately form and escape from the liquid. After the effervescence has ceased, a taste of the liquid will show you that the lemon juice has lost its acid nature, and has acquired in exchange a salty taste. Baking soda, when treated with an acid, is transformed into carbon dioxide and a salt. The various baking powders on the market to-day consist of baking soda and some acid substance, which acts upon the soda, forces it to give up its gas, and at the same time unites with the residue to form a harmless salt.

Cream of tartar contains sufficient acid to act on baking soda, and is a convenient and safe ingredient for baking powder. When soda and cream of tartar are mixed dry, they do not react on each other, neither do they combine rapidly in *cold* moist dough, but as soon as the heat of the oven penetrates the doughy mass, the cream of tartar combines with the soda and sets free the gas needed to raise the dough. The gas expands with the heat of the oven, raising the dough still more. Meanwhile, the dough itself is influenced by the heat and is stiffened to such an extent that it retains its inflated shape and spongy nature.

Many housewives look askance at ready-made baking powders and prefer to bake with soda and sour milk, soda and buttermilk, or soda and cream of tartar. Sour milk and buttermilk are quite as good as cream of tartar, because the lactic acid which they contain combines with the soda and liberates carbon dioxide, and forms a harmless residue in the dough.

CHAPTER XX 165

The desire of manufacturers to produce cheap baking powders led to the use of cheap acids and alkalies, regardless of the character of the resulting salt. Alum and soda were popular for some time; but careful examination proved that the particular salt produced by this combination was not readily absorbed by the stomach, and that its retention there was injurious to health. For this reason, many states have prohibited the use of alum in baking powders.

It is not only important to choose the ingredients carefully; it is also necessary to calculate the respective quantities of each, otherwise there will be an excess of acid or alkali for the stomach to take care of. A standard powder contains twice as much cream of tartar as of bicarbonate of soda, and the thrifty housewife who wishes to economize, can make for herself, at small cost, as good a baking powder as any on the market, by mixing tartar and soda in the above proportions and adding a little corn starch to keep the mixture dry.

The self-raising flour, so widely advertised by grocers, is flour in which these ingredients or their equivalent have been mixed by the manufacturer.

212. Soda Mints. Bicarbonate of soda is practically the sole ingredient of the soda mints popularly sold for indigestion. These correct a tendency to sour stomach because they counteract the surplus acid in the stomach, and form with it a safe neutral substance.

Seidlitz powder is a simple remedy consisting of two powders, one containing bicarbonate of soda, and the other, some acid such as cream of tartar. When these substances are dissolved in water and mixed, effervescence occurs, carbon dioxide escapes, and a solution of Rochelle salt remains.

212a. Source of Soda. An enormous quantity of sodium carbonate, or soda, as it is usually called, is needed in the manufacture of glass, soap, bleaching powders, and other commercial products. Formerly, the supply of soda was very limited because man was dependent upon natural deposits and upon ashes of sea plants for it. Common salt, sodium chloride, is abundant, and in 1775 a prize was offered to any one who would find a way to obtain soda from salt. As a result of this, soda was soon manufactured from common salt. In the most recent methods of manufacture, salt, water, ammonia, and carbon dioxide are made to react. Baking soda is formed from the reaction. The baking soda is then heated and decomposed into washing soda or the soda of commerce.

CHAPTER XXI

FERMENTATION

213. While baking powder is universally used for biscuits and cake, it is seldom, if ever, used for bread, because it does not furnish sufficient gas to lighten the tough heavy mass of bread dough. Then, too, most people prefer the taste of yeast-raised bread. There is a reason for this widespread preference, but to understand it, we must go somewhat far afield, and must study not only the bread of to-day, but the bread of antiquity, and the wines as well.

If grapes are crushed, they yield a liquid which tastes like the grapes; but if the liquid is allowed to stand in a warm place, it loses its original character, and begins to ferment, becoming, in the course of a few weeks, a strongly intoxicating drink. This is true not only of grape juice but also of the juice of all other sweet fruits; apple juice ferments to cider, currant juice to currant wine, etc. This phenomenon of fermentation is known to practically all races of men, and there is scarcely a savage tribe without some kind of fermented drink; in the tropics the fermented juice of the palm tree serves for wine; in the desert regions, the fermented juice of the century plant; and in still other regions, the root of the ginger plant is pressed into service.

The fermentation which occurs in bread making is similar to that which is responsible for the transformation of plant juices into intoxicating drinks. The former process is not so old, however, since the use of alcoholic beverages dates back to the very dawn of history, and the authentic record of raised or leavened bread is but little more than 3000 years old.

CHAPTER XXI 167

214. The Bread of Antiquity. The original method of bread making and the method employed by savage tribes of to-day is to mix crushed grain and water until a paste is formed, and then to bake this over a camp fire. The result is a hard compact substance known as unleavened bread. A considerable improvement over this tasteless mass is self-raised bread. If dough is left standing in a warm place a number of hours, it swells up with gas and becomes porous, and when baked, is less compact and hard than the savage bread. Exposure to air and warmth brings about changes in dough as well as in fruit juices, and alters the character of the dough and the bread made from it. Bread made in this way would not seem palatable to civilized man of the present day, accustomed, as he is, to delicious bread made light and porous by yeast; but to the ancients, the least softening and lightening was welcome, and self-fermented bread, therefore, supplanted the original unleavened bread.

Soon it was discovered that a pinch of this fermented dough acted as a starter on a fresh batch of dough. Hence, a little of the fermented dough was carefully saved from a batch, and when the next bread was made, the fermented dough, or leaven, was worked into the fresh dough and served to raise the mass more quickly and effectively than mere exposure to air and warmth could do in the same length of time. This use of leaven for raising bread has been practiced for ages.

Grape juice mixed with millet ferments quickly and strongly, and the Romans learned to use this mixture for bread raising, kneading a very small amount of it through the dough.

215. The Cause of Fermentation. Although alcoholic fermentation, and the fermentation which goes on in raising dough, were known and utilized for many years, the cause of the phenomenon was a sealed book until the nineteenth century. About that time it was discovered, through the use of the microscope, that fermenting liquids contain an army of minute plant organisms which not only live there, but which actually grow and multiply within the liquid. For growth and multiplication, food is necessary, and this the tiny plants get in abundance from the fruit juices; they feed upon the sugary matter and as they feed, they ferment it, changing it into carbon dioxide and alcohol. The carbon dioxide, in the form of small bubbles, passes off

CHAPTER XXI 168

from the fermenting mass, while the alcohol remains in the liquid, giving the stimulating effect desired by imbibers of alcoholic drinks. The unknown strange organisms were called yeast, and they were the starting point of the yeast cakes and yeast brews manufactured to-day on a large scale, not only for bread making but for the commercial production of beer, ale, porter, and other intoxicating drinks.

The grains, rye, corn, rice, wheat, from which meal is made, contain only a small quantity of sugar, but, on the other hand, they contain a large quantity of starch which is easily convertible into sugar. Upon this the tiny yeast plants in the dough feed, and, as in the case of the wines, ferment the sugar, producing carbon dioxide and alcohol. The dough is thick and sticky and the gas bubbles expand it into a spongy mass. The tiny yeast plants multiply and continue to make alcohol and gas, and in consequence, the dough becomes lighter and lighter. When it has risen sufficiently, it is kneaded and placed in an oven; the heat of the oven soon kills the yeast plants and drives the alcohol out of the bread; at the same time it expands the imprisoned gas bubbles and causes them to lighten and swell the bread still more. Meanwhile, the dough has become stiff enough to support itself. The result of the fermentation is a light, spongy loaf.

216. Where does Yeast come From? The microscopic plants which we call yeast are widely distributed in the air, and float around there until chance brings them in contact with a substance favorable to their growth, such as fruit juices and moist warm batter. Under the favorable conditions of abundant moisture, heat, and food, they grow and multiply rapidly, and cause the phenomenon of fermentation. Wild yeast settles on the skin of grapes and apples, but since it does not have access to the fruit juices within, it remains inactive very much as a seed does before it is planted. But when the fruit is crushed, the yeast plants get into the juice, and feeding on it, grow and multiply. The stray yeast plants which get into the sirup are relatively few, and hence fermentation is slow; it requires several weeks for currant wine to ferment, and several months for the juice of grapes to be converted into wine.

Stray yeast finds a favorable soil for growth in the warmth and moisture of a batter; but although the number of these stray plants is very large, it is insufficient to cause rapid fermentation, and if we depended upon wild yeast for bread raising, the result would not be to our liking.

When our remote ancestors saved a pinch of dough as leaven for the next baking, they were actually cultivating yeast, although they did not know it. The reserved portion served as a favorable breeding place to the yeast plants within it; they grew and reproduced amazingly, and became so numerous, that the small mass of old dough in which they were gathered served to leaven the entire batch at the next baking.

As soon as man learned that yeast plants caused fermentation in liquors and bread, he realized that it would be to his advantage to cultivate yeast and to add it to bread and to plant juices rather than to depend upon accidental and slow fermentation from wild yeast. Shortly after the discovery of yeast in the nineteenth century, man commenced his attempt to cultivate the tiny organisms. Their microscopic size added greatly to his trouble, and it was only after years of careful and tedious investigation that he was able to perfect the commercial yeast cakes and yeast brews universally used by bakers and brewers. The well-known compressed yeast cake is simply a mass of live and vigorous yeast plants, embedded in a soft, soggy material, and ready to grow and multiply as soon as they are placed under proper conditions of heat, moisture, and food. Seeds which remain on our shelves do not germinate, but those which are planted in the soil do; so it is with the yeast plants. While in the cake they are as lifeless as the seed; when placed in dough, or fruit juice, or grain water, they grow and multiply and cause fermentation.

CHAPTER XXII

BLEACHING

217. The beauty and the commercial value of uncolored fabrics depend upon the purity and perfection of their whiteness; a man's white collar and a woman's white waist must be pure white, without the slightest tinge of color. But all natural fabrics, whether they come from

is restricted by its combination with the other substances. By experiment it has been found that the addition to the bleaching solution of an acid, such as vinegar or lemon juice or sulphuric acid, causes the liberation of the chlorine. The chlorine thus set free reacts with the water and liberates oxygen; this in turn destroys the coloring matter in the fibers, and transforms the material into a bleached product.

The acid used to liberate the chlorine from the bleaching powder, and the chlorine also, rot materials with which they remain in contact for any length of time. For this reason, fabrics should be removed from the bleaching solution as soon as possible, and should then be rinsed in some solution, such as ammonia, which is capable of neutralizing the harmful substances; finally the fabric should be thoroughly rinsed in water in order that all foreign matter may be removed. The reason home bleaching is so seldom satisfactory is that most amateurs fail to realize the necessity of immediate neutralization and rinsing, and allow the fabric to remain too long in the bleaching solution, and allow it to dry with traces of the bleaching substances present in the fibers. Material treated in this way is thoroughly bleached, but is at the same time rotten and worthless. Chloride of lime is frequently used in laundry work; the clothes are whiter than when cleaned with soap and simple washing powders, but they soon wear out unless the precaution has been taken to add an "antichlor" or neutralizer to the bleaching solution.

220. Commercial Bleaching. In commercial bleaching the material to be bleached is first moistened with a very weak solution of sulphuric acid or hydrochloric acid, and is then immersed in the bleaching powder solution. As the moist material is drawn through the bleaching solution, the acid on the fabric acts upon the solution and releases chlorine. The chlorine liberates oxygen from the water. The oxygen in turn attacks the coloring matter and destroys it.

[Illustration: FIG. 159.--The material to be bleached is drawn through an acid *a*, then through a bleaching solution *b*, and finally through a neutralizing solution *c*.]

The bleached material is then immersed in a neutralizing bath and is finally rinsed thoroughly in water. Strips of cotton or linen many miles plants, like cotton and linen, or from animals, like wool and silk, contain more or less coloring matter, which impairs the whiteness. This coloring not only detracts from the appearance of fabrics which are to be worn uncolored, but it seriously interferes with the action of dyes, and at times plays the dyer strange tricks.

Natural fibers, moreover, are difficult to spin and weave unless some softening material such as wax or resin is rubbed lightly over them. The matter added to facilitate spinning and weaving generally detracts from the appearance of the uncolored fabric, and also interferes with successful dyeing. Thus it is easy to see that the natural coloring matter and the added foreign matter must be entirely removed from fabrics destined for commercial use. Exceptions to this general fact are sometimes made, because unbleached material is cheaper and more durable than the bleached product, and for some purposes is entirely satisfactory; unbleached cheesecloth and sheeting are frequently purchased in place of the more expensive bleached material. Formerly, the only bleaching agent known was the sun's rays, and linen and cotton were put out to sun for a week; that is, the unbleached fabrics were spread on the grass and exposed to the bleaching action of sun and dew.

[Illustration: FIG. 158.--Preparing chlorine from hydrochloric acid and manganese dioxide.]

218. An Artificial Bleaching Agent. While the sun's rays are effective as a bleaching agent, the process is slow; moreover, it would be impossible to expose to the sun's rays the vast quantity of fabrics used in the civilized world of to-day, and the huge and numerous bolts of material which daily come from our looms and factories must therefore be whitened by artificial means. The substance almost universally used as a rapid artificial bleaching agent is chlorine, best known to us as a constituent of common salt. Chlorine is never free in nature, but is found in combination with other substances, as, for example, in combination with sodium in salt, or with hydrogen in hydrochloric acid.

The best laboratory method of securing free chlorine is to heat in a water bath a mixture of hydrochloric acid and manganese dioxide, a compound containing one part of manganese and two parts of oxygen.

The heat causes the manganese dioxide to give up its oxygen immediately combines with the hydrogen of the hydrochloric a forms water. The manganese itself combines with part of the originally in the acid, but not with all. There is thus some free left over from the acid, and this passes off as a gas and can b collected, as in Figure 158. Free chlorine is heavier than air, hence when it leaves the exit tube it settles at the bottom of t displacing the air, and finally filling the bottle.

Chlorine is a very active substance and combines readily wit substances, but especially with hydrogen; if chlorine comes i with steam, it abstracts the hydrogen and unites with it to for hydrochloric acid, but it leaves the oxygen free and uncombi tendency of chlorine to combine with hydrogen makes it valu bleaching agent. In order to test the efficiency of chlorine as bleaching agent, drop a wet piece of colored gingham or cal bottle of chlorine, and notice the rapid disappearance of col sample. If unbleached muslin is used, the moist strip loses i yellowish hue and becomes a clear, pure white. The explan bleaching power of chlorine is that the chlorine combines wi hydrogen of the water and sets oxygen free; the uncombine oxygen oxidizes the coloring matter in the cloth and destroy

Chlorine has no effect on dry material, as may be seen if w gingham into the jar; in this case there is no water to furnisl for combination with the chlorine, and no oxygen to be set 1

219. Bleaching Powder. Chlorine gas has a very injurious e human body, and hence cannot be used directly as a blead It attacks the mucous membrane of the nose and lungs, ar the effect of a severe cold or catarrh, and when inhaled, ca But certain compounds of chlorine are harmless, and can l instead of chlorine for destroying either natural or artificial these compounds, namely, chloride of lime, is the almost u bleaching agent of commerce. It comes in the form of pow can be dissolved in water to form the bleaching solution in colored fabrics are immersed. But fabrics immersed in a b powder solution do not lose their color as would naturally The reason for this is that the chlorine gas is not free to d

CHAPTER XXII 173

long are drawn by machinery into and out of the various solutions (Fig. 159), are then passed over pressing rollers, and emerge snow white, ready to be dyed or to be used as white fabric.

221. Wool and Silk Bleaching. Animal fibers like silk, wool, and feathers, and some vegetable fibers like straw, cannot be bleached by means of chlorine, because it attacks not only the coloring matter but the fiber itself, and leaves it shrunken and inferior. Cotton and linen fibers, apart from the small amount of coloring matter present in them, contain nothing but carbon, oxygen, and hydrogen, while animal fibers contain in addition to these elements some compounds of nitrogen. The presence of these nitrogen compounds influences the action of the chlorine and produces unsatisfactory results. For animal fibers it is therefore necessary to discard chlorine as a bleaching agent, and to substitute a substance which will have a less disastrous action upon the fibers. Such a substance is to be had in sulphurous acid. When sulphur burns, as in a match, it gives off disagreeable fumes, and if these are made to bubble into a vessel containing water, they dissolve and form with the water a substance known as sulphurous acid. That this solution has bleaching properties is shown by the fact that a colored cloth dipped into it loses its color, and unbleached fabrics immersed in it are whitened. The harmless nature of sulphurous acid makes it very desirable as a bleaching agent, especially in the home.

Silk, lace, and wool when bleached with chlorine become hard and brittle, but when whitened with sulphurous acid, they retain their natural characteristics.

This mild form of a bleaching substance has been put to uses which are now prohibited by the pure food laws. In some canneries common corn is whitened with sulphurous acid, and is then sold under false representations. Cherries are sometimes bleached and then colored with the bright shades which under natural conditions indicate freshness.

Bleaching with chlorine is permanent, the dyestuff being destroyed by the chlorine; but bleaching with sulphurous acid is temporary, because the milder bleach does not actually destroy the dyestuff, but merely modifies it, and in time the natural yellow color of straw, cotton, and

CHAPTER XXII

linen reappears. The yellowing of straw hats during the summer is familiar to everyone; the straw is merely resuming its natural color which had been modified by the sulphurous acid solution applied to the straw when woven.

222. Why the Color Returns. Some of the compounds formed by the sulphurous acid bleaching process are gradually decomposed by sunlight, and in consequence the original color is in time partially restored. The portion of a hat protected by the band retains its fresh appearance because the light has not had access to it. Silks and other fine fabrics bleached in this way fade with age, and assume an unnatural color. One reason for this is that the dye used to color the fabric requires a clear white background, and loses its characteristic hues when its foundation is yellow instead of white. Then, too, dyestuffs are themselves more or less affected by light, and fade slowly under a strong illumination.

Materials which are not exposed directly to an intense and prolonged illumination retain their whiteness for a long time, and hence dress materials and hats which have been bleached with sulphurous acid should be protected from the sun's glare when not in use.

223. The Removal of Stains. Bleaching powder is very useful in the removal of stains from white fabrics. Ink spots rubbed with lemon juice and dipped in bleaching solution fade away and leave on the cloth no trace of discoloration. Sometimes these stains can be removed by soaking in milk, and where this is possible, it is the better method.

Bleaching solution, however, while valuable in the removal of some stains, is unable to remove paint stains, because paints owe their color to mineral matter, and on this chlorine is powerless to act. Paint stains are best removed by the application of gasoline followed by soap and water.

very large, it is insufficient to cause rapid fermentation, and if we depended upon wild yeast for bread raising, the result would not be to our liking.

When our remote ancestors saved a pinch of dough as leaven for the next baking, they were actually cultivating yeast, although they did not know it. The reserved portion served as a favorable breeding place to the yeast plants within it; they grew and reproduced amazingly, and became so numerous, that the small mass of old dough in which they were gathered served to leaven the entire batch at the next baking.

As soon as man learned that yeast plants caused fermentation in liquors and bread, he realized that it would be to his advantage to cultivate yeast and to add it to bread and to plant juices rather than to depend upon accidental and slow fermentation from wild yeast. Shortly after the discovery of yeast in the nineteenth century, man commenced his attempt to cultivate the tiny organisms. Their microscopic size added greatly to his trouble, and it was only after years of careful and tedious investigation that he was able to perfect the commercial yeast cakes and yeast brews universally used by bakers and brewers. The well-known compressed yeast cake is simply a mass of live and vigorous yeast plants, embedded in a soft, soggy material, and ready to grow and multiply as soon as they are placed under proper conditions of heat, moisture, and food. Seeds which remain on our shelves do not germinate, but those which are planted in the soil do; so it is with the yeast plants. While in the cake they are as lifeless as the seed; when placed in dough, or fruit juice, or grain water, they grow and multiply and cause fermentation.

CHAPTER XXII

BLEACHING

217. The beauty and the commercial value of uncolored fabrics depend upon the purity and perfection of their whiteness; a man's white collar and a woman's white waist must be pure white, without the slightest tinge of color. But all natural fabrics, whether they come from

CHAPTER XXII 170

plants, like cotton and linen, or from animals, like wool and silk, contain more or less coloring matter, which impairs the whiteness. This coloring not only detracts from the appearance of fabrics which are to be worn uncolored, but it seriously interferes with the action of dyes, and at times plays the dyer strange tricks.

Natural fibers, moreover, are difficult to spin and weave unless some softening material such as wax or resin is rubbed lightly over them. The matter added to facilitate spinning and weaving generally detracts from the appearance of the uncolored fabric, and also interferes with successful dyeing. Thus it is easy to see that the natural coloring matter and the added foreign matter must be entirely removed from fabrics destined for commercial use. Exceptions to this general fact are sometimes made, because unbleached material is cheaper and more durable than the bleached product, and for some purposes is entirely satisfactory; unbleached cheesecloth and sheeting are frequently purchased in place of the more expensive bleached material. Formerly, the only bleaching agent known was the sun's rays, and linen and cotton were put out to sun for a week; that is, the unbleached fabrics were spread on the grass and exposed to the bleaching action of sun and dew.

[Illustration: FIG. 158.--Preparing chlorine from hydrochloric acid and manganese dioxide.]

218. **An Artificial Bleaching Agent.** While the sun's rays are effective as a bleaching agent, the process is slow; moreover, it would be impossible to expose to the sun's rays the vast quantity of fabrics used in the civilized world of to-day, and the huge and numerous bolts of material which daily come from our looms and factories must therefore be whitened by artificial means. The substance almost universally used as a rapid artificial bleaching agent is chlorine, best known to us as a constituent of common salt. Chlorine is never free in nature, but is found in combination with other substances, as, for example, in combination with sodium in salt, or with hydrogen in hydrochloric acid.

The best laboratory method of securing free chlorine is to heat in a water bath a mixture of hydrochloric acid and manganese dioxide, a compound containing one part of manganese and two parts of oxygen.

CHAPTER XXII

The heat causes the manganese dioxide to give up its oxygen, which immediately combines with the hydrogen of the hydrochloric acid and forms water. The manganese itself combines with part of the chlorine originally in the acid, but not with all. There is thus some free chlorine left over from the acid, and this passes off as a gas and can be collected, as in Figure 158. Free chlorine is heavier than air, and hence when it leaves the exit tube it settles at the bottom of the jar, displacing the air, and finally filling the bottle.

Chlorine is a very active substance and combines readily with most substances, but especially with hydrogen; if chlorine comes in contact with steam, it abstracts the hydrogen and unites with it to form hydrochloric acid, but it leaves the oxygen free and uncombined. This tendency of chlorine to combine with hydrogen makes it valuable as a bleaching agent. In order to test the efficiency of chlorine as a bleaching agent, drop a wet piece of colored gingham or calico into the bottle of chlorine, and notice the rapid disappearance of color from the sample. If unbleached muslin is used, the moist strip loses its natural yellowish hue and becomes a clear, pure white. The explanation of the bleaching power of chlorine is that the chlorine combines with the hydrogen of the water and sets oxygen free; the uncombined free oxygen oxidizes the coloring matter in the cloth and destroys it.

Chlorine has no effect on dry material, as may be seen if we put dry gingham into the jar; in this case there is no water to furnish hydrogen for combination with the chlorine, and no oxygen to be set free.

219. Bleaching Powder. Chlorine gas has a very injurious effect on the human body, and hence cannot be used directly as a bleaching agent. It attacks the mucous membrane of the nose and lungs, and produces the effect of a severe cold or catarrh, and when inhaled, causes death. But certain compounds of chlorine are harmless, and can be used instead of chlorine for destroying either natural or artificial dyes. One of these compounds, namely, chloride of lime, is the almost universal bleaching agent of commerce. It comes in the form of powder, which can be dissolved in water to form the bleaching solution in which the colored fabrics are immersed. But fabrics immersed in a bleaching powder solution do not lose their color as would naturally be expected. The reason for this is that the chlorine gas is not free to do its work, but

is restricted by its combination with the other substances. By experiment it has been found that the addition to the bleaching solution of an acid, such as vinegar or lemon juice or sulphuric acid, causes the liberation of the chlorine. The chlorine thus set free reacts with the water and liberates oxygen; this in turn destroys the coloring matter in the fibers, and transforms the material into a bleached product.

The acid used to liberate the chlorine from the bleaching powder, and the chlorine also, rot materials with which they remain in contact for any length of time. For this reason, fabrics should be removed from the bleaching solution as soon as possible, and should then be rinsed in some solution, such as ammonia, which is capable of neutralizing the harmful substances; finally the fabric should be thoroughly rinsed in water in order that all foreign matter may be removed. The reason home bleaching is so seldom satisfactory is that most amateurs fail to realize the necessity of immediate neutralization and rinsing, and allow the fabric to remain too long in the bleaching solution, and allow it to dry with traces of the bleaching substances present in the fibers. Material treated in this way is thoroughly bleached, but is at the same time rotten and worthless. Chloride of lime is frequently used in laundry work; the clothes are whiter than when cleaned with soap and simple washing powders, but they soon wear out unless the precaution has been taken to add an "antichlor" or neutralizer to the bleaching solution.

220. Commercial Bleaching. In commercial bleaching the material to be bleached is first moistened with a very weak solution of sulphuric acid or hydrochloric acid, and is then immersed in the bleaching powder solution. As the moist material is drawn through the bleaching solution, the acid on the fabric acts upon the solution and releases chlorine. The chlorine liberates oxygen from the water. The oxygen in turn attacks the coloring matter and destroys it.

[Illustration: FIG. 159.--The material to be bleached is drawn through an acid *a*, then through a bleaching solution *b*, and finally through a neutralizing solution *c*.]

The bleached material is then immersed in a neutralizing bath and is finally rinsed thoroughly in water. Strips of cotton or linen many miles

CHAPTER XXII 173

long are drawn by machinery into and out of the various solutions (Fig. 159), are then passed over pressing rollers, and emerge snow white, ready to be dyed or to be used as white fabric.

221. Wool and Silk Bleaching. Animal fibers like silk, wool, and feathers, and some vegetable fibers like straw, cannot be bleached by means of chlorine, because it attacks not only the coloring matter but the fiber itself, and leaves it shrunken and inferior. Cotton and linen fibers, apart from the small amount of coloring matter present in them, contain nothing but carbon, oxygen, and hydrogen, while animal fibers contain in addition to these elements some compounds of nitrogen. The presence of these nitrogen compounds influences the action of the chlorine and produces unsatisfactory results. For animal fibers it is therefore necessary to discard chlorine as a bleaching agent, and to substitute a substance which will have a less disastrous action upon the fibers. Such a substance is to be had in sulphurous acid. When sulphur burns, as in a match, it gives off disagreeable fumes, and if these are made to bubble into a vessel containing water, they dissolve and form with the water a substance known as sulphurous acid. That this solution has bleaching properties is shown by the fact that a colored cloth dipped into it loses its color, and unbleached fabrics immersed in it are whitened. The harmless nature of sulphurous acid makes it very desirable as a bleaching agent, especially in the home.

Silk, lace, and wool when bleached with chlorine become hard and brittle, but when whitened with sulphurous acid, they retain their natural characteristics.

This mild form of a bleaching substance has been put to uses which are now prohibited by the pure food laws. In some canneries common corn is whitened with sulphurous acid, and is then sold under false representations. Cherries are sometimes bleached and then colored with the bright shades which under natural conditions indicate freshness.

Bleaching with chlorine is permanent, the dyestuff being destroyed by the chlorine; but bleaching with sulphurous acid is temporary, because the milder bleach does not actually destroy the dyestuff, but merely modifies it, and in time the natural yellow color of straw, cotton, and linen reappears. The yellowing of straw hats during the summer is familiar to everyone; the straw is merely resuming its natural color which had been modified by the sulphurous acid solution applied to the straw when woven.

222. Why the Color Returns. Some of the compounds formed by the sulphurous acid bleaching process are gradually decomposed by sunlight, and in consequence the original color is in time partially restored. The portion of a hat protected by the band retains its fresh appearance because the light has not had access to it. Silks and other fine fabrics bleached in this way fade with age, and assume an unnatural color. One reason for this is that the dye used to color the fabric requires a clear white background, and loses its characteristic hues when its foundation is yellow instead of white. Then, too, dyestuffs are themselves more or less affected by light, and fade slowly under a strong illumination.

Materials which are not exposed directly to an intense and prolonged illumination retain their whiteness for a long time, and hence dress materials and hats which have been bleached with sulphurous acid should be protected from the sun's glare when not in use.

223. The Removal of Stains. Bleaching powder is very useful in the removal of stains from white fabrics. Ink spots rubbed with lemon juice and dipped in bleaching solution fade away and leave on the cloth no trace of discoloration. Sometimes these stains can be removed by soaking in milk, and where this is possible, it is the better method.

Bleaching solution, however, while valuable in the removal of some stains, is unable to remove paint stains, because paints owe their color to mineral matter, and on this chlorine is powerless to act. Paint stains are best removed by the application of gasoline followed by soap and water.

CHAPTER XXIII

DYEING

224. Dyes. One of the most important and lucrative industrial processes of the world to-day is that of staining and dyeing. Whether we consider the innumerable shades of leather used in shoes and harnesses and upholstery; the multitude of colors in the paper which covers our walls and reflects light ranging from the somber to the gay, and from the delicate to the gorgeous; the artificial scenery which adorns the stage and by its imitation of trees and flowers and sky translates us to the Forest of Arden; or whether we consider the uncounted varieties of color in dress materials, in carpets, and in hangings, we are dealing with substances which owe their beauty to dyes and dyestuffs.

The coloring of textile fabrics, such as cotton, wool, and silk, far outranks in amount and importance that of leather, paper, etc., and hence the former only will be considered here; but the theories and facts relative to textile dyeing are applicable in a general way to all other forms as well.

225. Plants as a Source of Dyes. Among the most beautiful examples of man's handiwork are the baskets and blankets of the North American Indians, woven with a skill which cannot be equaled by manufacturers, and dyed in mellow colors with a few simple dyes extracted from local plants. The magnificent rugs and tapestries of Persia and Turkey, and the silks of India and Japan, give evidence that a knowledge of dyes is widespread and ancient. Until recently, the vegetable world was the source of practically all coloring matter, the pulverized root of the madder plant yielding the reds, the leaves and stems of the indigo plant the blues, the heartwood of the tropical logwood tree the blacks and grays, and the fruit of certain palm and locust trees yielding the soft browns. So great was the commercial demand for dyestuffs that large areas of land were given over to the exclusive cultivation of the more important dye plants. Vegetable dyes are now, however, rarely used because about the year 1856 it was discovered that dyes could be obtained from coal tar, the thick sticky liquid formed as a by-product in the manufacture of coal gas. These

CHAPTER XXIII

artificial coal-tar, or aniline, dyes have practically undisputed sway to-day, and the vast areas of land formerly used for the cultivation of vegetable dyes are now free for other purposes.

226. Wool and Cotton Dyeing. If a piece of wool is soaked in a solution of a coal-tar dye, such as magenta, the fiber of the cloth draws some of the dye out of the solution and absorbs it, becoming in consequence beautifully colored. The coloring matter becomes "part and parcel," as it were, of the wool fiber, because repeated washing of the fabric fails to remove the newly acquired color; the magenta coloring matter unites chemically with the fiber of the wool, and forms with it a compound insoluble in water, and hence fast to washing.

But if cotton is used instead of wool, the acquired color is very faint, and washes off readily. This is because cotton fibers possess no chemical substance capable of uniting with the coloring matter to form a compound insoluble in water.

If magenta is replaced by other artificial dyes,--for example, scarlets,--the result is similar; in general, wool material absorbs dye readily, and uniting with it is permanently dyed. Cotton material, on the other hand, does not combine chemically with coloring matter and therefore is only faintly tinged with color, and loses this when washed. When silk and linen are tested, it is found that the former behaves in a general way as did wool, while the linen has more similarity to the cotton. That vegetable fibers, such as cotton and linen, should act differently toward coloring matter from animal fibers, such as silk and wool, is not surprising when we consider that the chemical nature of the two groups is very different; vegetable fibers contain only oxygen, carbon, and hydrogen, while animal fibers always contain nitrogen in addition, and in many cases sulphur as well.

227. The Selection of Dyes. When silk and wool, cotton and linen, are tested in various dye solutions, it is found that the former have, in general, a great affinity for coloring matter and acquire a permanent color, but that cotton and linen, on the other hand, have little affinity for dyestuffs. The color acquired by vegetable fibers is, therefore, usually faint.

CHAPTER XXIII 177

There are, of course, many exceptions to the general statement that animal fibers dye readily and vegetable fibers poorly, because certain dyes fail utterly with woolen and silk material and yet are fairly satisfactory when applied to cotton and linen fabrics. Then, too, a dye which will color silk may not have any effect on wool in spite of the fact that wool, like silk, is an animal fiber; and certain dyestuffs to which cotton responds most beautifully are absolutely without effect on linen.

The nature of the material to be dyed determines the coloring matter to be used; in dyeing establishments a careful examination is made of all textiles received for dyeing, and the particular dyestuffs are then applied which long experience has shown to be best suited to the material in question. Where "mixed goods," such as silk and wool, or cotton and wool, are concerned, the problem is a difficult one, and the countless varieties of gorgeously colored mixed materials give evidence of high perfection in the art of dyeing and weaving.

Housewives who wish to do successful home dyeing should therefore not purchase dyes indiscriminately, but should select the kind best suited to the material, because the coloring principle which will remake a silk waist may utterly ruin a woolen skirt or a linen suit. Powders designed for special purposes may be purchased from druggists.

228. Indirect Dyeing. We have seen that it is practically impossible to color cotton and linen in a simple manner with any degree of permanency, because of the lack of chemical action between vegetable fibers and coloring matter. But the varied uses to which dyed articles are put make fastness of color absolutely necessary. A shirt, for example, must not be discolored by perspiration, nor a waist faded by washing, nor a carpet dulled by sweeping with a dampened broom. In order to insure permanency of dyes, an indirect method was originated which consisted of adding to the fibers a chemical capable of acting upon the dye and forming with it a colored compound insoluble in water, and hence "safe." For example, cotton material dyed directly in logwood solution has almost no value, but if it is soaked in a solution of oxalic acid and alum until it becomes saturated with the chemicals, and is then transferred to a logwood bath, the color acquired is fast and beautiful.

CHAPTER XXIII

This method of indirect dyeing is known as the mordanting process; it consists of saturating the fabric to be dyed with chemicals which will unite with the coloring matter to form compounds unaffected by water. The chemicals are called mordants.

229. How Variety of Color is Secured. The color which is fixed on the fabric as a result of chemical action between mordant and dye is frequently very different from that of the dye itself. Logwood dye when used alone produces a reddish brown color of no value either for beauty or permanence; but if the fabric to be dyed is first mordanted with a solution of alum and oxalic acid and is then immersed in a logwood bath, it acquires a beautiful blue color.

Moreover, since the color acquired depends upon the mordant as well as upon the dye, it is often possible to obtain a wide range of colors by varying the mordant used, the dye remaining the same. For example, with alum and oxalic acid as a mordant and logwood as a dye, blue is obtained; but with a mordant of ferric sulphate and a dye of logwood, blacks and grays result. Fabrics immersed directly in alizarin acquire a reddish yellow tint; when, however, they are mordanted with certain aluminium compounds they acquire a brilliant Turkey red, when mordanted with chromium compounds, a maroon, and when mordanted with iron compounds, the various shades of purple, lilac, and violet result.

230. Color Designs in Cloth. It is thought that the earliest attempts at making "fancy materials" consisted in painting designs on a fabric by means of a brush. In more recent times the design was cut in relief on hard wood, the relief being then daubed with coloring matter and applied by hand to successive portions of the cloth. The most modern method of design-making is that of machine or roller printing. In this, the relief blocks are replaced by engraved copper rolls which rotate continuously and in the course of their rotation automatically receive coloring matter on the engraved portion. The cloth is to be printed is then drawn uniformly over the rotating roll, receiving color from the engraved design; in this way, the color pattern is automatically printed on the cloth with perfect regularity. In cases where the fabrics do not unite directly with the coloring matter, the design is supplied with a mordant and the impression made on the fabric is that of the mordant;

when the fabric is later transferred to a dye bath, the mordanted portions, represented by the design, unite with the coloring matter and thus form the desired color patterns.

Unless the printing is well done, the coloring matter does not thoroughly penetrate the material, and only a faint blurred design appears on the back of the cloth; the gaudy designs of cheap calicoes and ginghams often do not show at all on the under side. Such carelessly made prints are not fast to washing or light, and soon fade. But in the better grades of material the printing is well done, and the color designs are fairly fast, and a little care in the laundry suffices to eliminate any danger of fading.

Color designs of the greatest durability are produced by the weaving together of colored yarns. When yarn is dyed, the coloring matter penetrates to every part of the fiber, and hence the patterns formed by the weaving together of well-dyed yarns are very fast to light and water.

If the color designs to be woven in the cloth are intricate, complex machinery is necessary and skillful handwork; hence, patterns formed by the weaving of colored yarns are expensive and less common than printed fabrics.

CHAPTER XXIV

CHEMICALS AS DISINFECTANTS AND PRESERVATIVES

231. The prevention of disease epidemics is one of the most striking achievements of modern science. Food, clothing, furniture, and other objects contaminated in any way by disease germs may be disinfected by chemicals or by heat, and widespread infection from persons suffering with a contagious disease may be prevented.

[Illustration: FIG. 160.--Pasteurizing apparatus, an arrangement by which milk is conveniently heated to destroy disease germs.]

When disease germs are within the body, the problem is far from simple, because chemicals which would effectively destroy the germs would be fatal to life itself. But when germs are outside the body, as in water or milk, or on clothing, dishes, or furniture, they can be easily killed. One of the best methods of destroying germs is to subject them to intense heat. Contaminated water is made safe by boiling for a few minutes, because the strong heat destroys the disease-producing germs. Scalded or Pasteurized milk saves the lives of scores of babies, because the germs of summer complaint which lurk in poor milk are killed and rendered harmless in the process of scalding. Dishes used by consumptives, and persons suffering from contagious diseases, can be made harmless by thorough washing in thick suds of almost boiling water.

The bedding and clothing of persons suffering with diphtheria, tuberculosis, and other germ diseases should always be boiled and hung to dry in the bright sunlight. Heat and sunshine are two of the best disinfectants.

232. Chemicals. Objects, such as furniture, which cannot be boiled, are disinfected by the use of any one of several chemicals, such as sulphur, carbolic acid, chloride of lime, corrosive sublimate, etc.

One of the simplest methods of disinfecting consists in burning sulphur in a room whose doors, windows, and keyholes have been closed, so that the burning fumes cannot escape, but remain in the room long enough to destroy disease germs. This is probably the most common means of fumigation.

For general purposes, carbolic acid is one of the very best disinfectants, but must be used with caution, as it is a deadly poison except when very dilute.

Chloride of lime when exposed to the air and moisture slowly gives off chlorine, and can be used as a disinfectant because the gas thus set free attacks germs and destroys them. For this reason chloride of lime is an excellent disinfectant of drainpipes. Certain bowel troubles, such as diarrhoea, are due to microbes, and if the waste matter of a person suffering from this or similar diseases is allowed passage through the

CHAPTER XXIV

drainage system, much damage may be done. But a small amount of chloride of lime in the closet bowl will insure disinfection.

233. Personal Disinfection. The hands may gather germs from any substances or objects with which they come in contact; hence the hands should be washed with soap and water, and especially before eating. Physicians who perform operations wash not only their hands, but their instruments, sterilizing the latter by placing them in boiling water for several minutes.

Cuts and wounds allow easy access to the body; a small cut has been known to cause death because of the bacteria which found their way into the open wound and produced disease. In order to destroy any germs which may have entered into the cut from the instrument, it is well to wash out the wound with some mild disinfectant, such as very dilute carbolic acid or hydrogen peroxide, and then to bind the wound with a clean cloth, to prevent later entrance of germs.

234. Chemicals as Food Preservatives. The spoiling of meats and soups, and the souring of milk and preserves, are due to germs which, like those producing disease, can be destroyed by heat and by chemicals.

Milk heated to the boiling point does not sour readily, and successful canning consists in cooking fruits and vegetables until all the germs are killed, and then sealing the cans so that germs from outside cannot find entrance and undo the work of the canner.

Some dealers and manufacturers have learned that certain chemicals will act as food preservatives, and hence they have replaced the safe method of careful canning by the quicker and simpler plan of adding chemicals to food. Catchup, sauces, and jellies are now frequently preserved in this way. But the chemicals which destroy bacteria frequently injure the consumer as well. And so much harm has been done by food preservatives that the pure food laws require that cans and bottles contain a labeled statement of the kind and quantity of chemicals used.

Even milk is not exempt, but is doctored to prevent souring, the preservative most generally used by milk dealers being formaldehyde. The vast quantity of milk consumed by young and old, sick and well, makes the use of formaldehyde a serious menace to health, because no constitution can endure the injury done by the constant use of preservatives.

The most popular and widely used preservatives of meats are borax and boric acid. These chemicals not only arrest decay, but partially restore to old and bad meat the appearance of freshness; in this way unscrupulous dealers are able to sell to the public in one form or other meats which may have undergone partial decomposition; sausage frequently contains partially decomposed meat, restored as it were by chemicals.

In jams and catchups there is abundant opportunity for preservatives; badly or partially decayed fruits are sometimes disinfected and used as the basis of foods sold by so-called good dealers. Benzoate of soda, and salicylic acid are the chemicals most widely employed for this purpose, with coal-tar dyes to simulate the natural color of the fruit.

Many of the cheap candies sold by street venders are not fit for consumption, since they are not only made of bad material, but are frequently in addition given a light dipping in varnish as a protection against the decaying influences of the atmosphere.

The only wise preservatives are those long known and employed by our ancestors; salt, vinegar, and spices are all food preservatives, but they are at the same time substances which in small amounts are not injurious to the body. Smoked herring and salted mackerel are chemically preserved foods, but they are none the less safe and digestible.

235. The Preservation of Wood and Metal. The decaying of wood and the rusting of metal are due to the action of air and moisture. When wood and metal are surrounded with a covering which neither air nor moisture can penetrate, decay and rust are prevented. Paint affords such a protective covering. The main constituent of paint is a compound of white lead or other metallic substance; this is mixed with

linseed oil or its equivalent in order that it may be spread over wood and metal in a thin, even coating. After the mixture has been applied, it hardens and forms a tough skin fairly impervious to weathering. For the sake of ornamentation, various colored pigments are added to the paint and give variety of effect.

Railroad ties and street paving blocks are ordinarily protected by oil rather than paint. Wood is soaked in creosote oil until it becomes thoroughly saturated with the oily substance. The pores of the wood are thus closed to the entrance of air and moisture, and decay is avoided. Wood treated in this way is very durable. Creosote is poisonous to insects and many small animals, and thus acts as a preservation not only against the elements but against animal life as well.

CHAPTER XXV

DRUGS AND PATENT MEDICINES

236. Stimulants and Narcotics. Man has learned not only the action of substances upon each other, such as bleaching solution upon coloring matter, washing soda upon grease, acids upon bases, but also the effect which certain chemicals have upon the human body.

Drugs and their varying effects upon the human system have been known to mankind from remote ages; in the early days, familiar leaves, roots, and twigs were steeped in water to form medicines which served for the treatment of all ailments. In more recent times, however, these simple herb teas have been supplanted by complex drugs, and now medicines are compounded not only from innumerable plant products, but from animal and mineral matter as well. Quinine, rhubarb, and arnica are examples of purely vegetable products; iron, mercury, and arsenic are equally well known as distinctly mineral products, while cod-liver oil is the most familiar illustration of an animal remedy. Ordinarily a combination of products best serves the ends of the physician.

CHAPTER XXV

Substances which, like cod-liver oil, serve as food to a worn-out body, or, like iron, tend to enrich the blood, or, like quinine, aid in bringing an abnormal system to a healthy condition, are valuable servants and cannot be entirely dispensed with so long as man is subject to disease.

But substances which, like opium, laudanum, and alcohol, are not required by the body as food, or as a systematic, intelligent aid to recovery, but are taken solely for the stimulus aroused or for the insensibility induced, are harmful to man, and cannot be indulged in by him without ultimate mental, moral, and physical loss. Substances of the latter class are known as narcotics and stimulants.

237. The Cost of Health. In the physical as in the financial world, nothing is to be had without a price. Vigor, endurance, and mental alertness are bought by hygienic living; that is, by proper food, fresh air, exercise, cleanliness, and reasonable hours. Some people wish vigor, endurance, etc., but are unwilling to live the life which will develop these qualities. Plenty of sleep, exercise, and simple food all tend to lay the foundations of health. Many, however, are not willing to take the care necessary for healthful living, because it would force them to sacrifice some of the hours of pleasure. Sooner or later, these pleasure-seekers begin to feel tired and worn, and some of them turn to drugs and narcotics for artificial strength. At first the drugs seem to restore the lost energy, and without harm; however, the cost soon proves to be one of the highest Nature ever demands.

238. The Uncounted Cost. The first and most obvious effect of opium, for example, is to deaden pain and to arouse pleasure; but while the drug is producing these soothing sensations, it interferes with bodily functions. Secretion, digestion, absorption of food, and the removal of waste matters are hindered. Continued use of the drug leads to headache, exhaustion, nervous depression, and heart weakness. There is thus a heavy toll reckoned against the user, and the creditor is relentless in demanding payment.

Moreover, the respite allowed by a narcotic is exceedingly brief, and a depression which is long and deep inevitably follows. In order to overcome this depression, recourse is usually had to a further dose, and as time goes on, the intervals of depression become more frequent and lasting, and the necessity to overcome them increases. Thus without intention one finds one's self bound to the drug, its fast victim. The sanatoria of our country are crowded with people who are trying to free themselves of a drug habit into which they have drifted unintentionally if not altogether unknowingly. What is true of opium is equally applicable to other narcotics.

239. **The Right Use of Narcotics.** In the hands of the physician, narcotics are a great blessing. In some cases, by relieving pain, they give the system the rest necessary for overcoming the cause of the pain. Only those who know of the suffering endured in former times can fully appreciate the decrease in pain brought about by the proper use of narcotics.

240. **Patent Medicines, Cough Sirups.** A reputable physician is solicitous regarding the permanent welfare of his patient and administers carefully chosen and harmless drugs. Mere medicine venders, however, ignore the good of mankind, and flood the market with cheap patent preparations which delude and injure those who purchase, but bring millions of dollars to those who manufacture.

Practically all of these patent, or proprietary, preparations contain a large proportion of narcotics or stimulants, and hence the benefit which they seem to afford the user is by no means genuine; examination shows that the relief brought by them is due either to a temporary deadening of sensibilities by narcotics or to a fleeting stimulation by alcohol and kindred substances.

Among the most common ailments of both young and old are coughs and colds; hence many patent cough mixtures have been manufactured and placed on the market for the consumption of a credulous public. Such "quick cures" almost invariably contain one or more narcotic drugs, and not only do not relieve the cold permanently, but occasion subsequent disorders. Even lozenges and pastilles are not free from fraud, but have a goodly proportion of narcotics, containing in some cases chloroform, morphine, and ether.

The widespread use of patent cough medicines is due largely to the fact that many persons avoid consulting a physician about so trivial an

CHAPTER XXV 186

ailment as an ordinary cold, or are reluctant to pay a medical fee for what seems a slight indisposition and hence attempt to doctor themselves.

Catarrh is a very prevalent disease in America, and consequently numerous catarrh remedies have been devised, most of which contain in a disguised form the pernicious drug, cocaine. Laws have been enacted which require on the labels a declaration of the contents of the preparation, both as to the kind of drug used and the amount, and the choice of accepting or refusing such mixtures is left to the individual. But the great mass of people are ignorant of the harmful nature of drugs in general, and hence do not even read the self-accusing label, or if they do glance at it, fail to comprehend the dangerous nature of the drugs specified there. In order to safeguard the uninformed purchaser and to restrict the manufacture of harmful patent remedies, some states limit the sale of all preparations containing narcotics and thus give free rein to neither consumer nor producer.

241. Soothing Sirups; Soft Drinks. The development of a race is limited by the mental and physical growth of its children, and yet thousands of its children are annually stunted and weakened by drugs, because most colic cures, teething concoctions, and soothing syrups are merely agreeably flavored drug mixtures. Those who have used such preparations freely, know that a child usually becomes fretful and irritable between doses, and can be quieted only by larger and more frequent supplies. A habit formed in this way is difficult to overcome, and many a child when scarcely over its babyhood had a craving which in later years may lead to systematic drug taking. And even though the pernicious drug craving is not created, considerable harm is done to the child, because its body is left weak and non-resistant to diseases of infancy and childhood.

Many of our soft drinks contain narcotics. The use of the coca leaf and the kola nut for such preparations has increased very greatly within the last few years, and doubtless legislation will soon be instituted against the indiscriminate sale of soft drinks.

242. Headache Powders. The stress and strain of modern life has opened wide the door to a multitude of bodily ills, among which may be

mentioned headache. Work must be done and business attended to, and the average sufferer does not take time from his vocation to investigate the cause of the headache, but unthinkingly grasps at any remedy which will remove the immediate pain, and utterly disregards later injury. The relief afforded by most headache mixtures is due to the presence of antipyrin or acetanilid, and it has been shown conclusively that these drugs weaken heart action, diminish circulation, reduce the number of red corpuscles in the blood, and bring on a condition of chronic anemia. Pallid cheeks and blue lips are visible evidence of the too frequent use of headache powders.

The labels required by law are often deceptive and convey no adequate idea of the amount of drug consumed; for example, 240 grains of acetanilid to an ounce seems a small quantity of drug for a powder, but when one considers that there are only 480 grains in an ounce, it will be seen that each powder is one half acetanilid.

Powders taken in small quantities and at rare intervals are apparently harmless; but they never remove the cause of the trouble, and hence the discomfort soon returns with renewed force. Ordinarily, hygienic living will eliminate the source of the trouble, but if it does not, a physician should be consulted and medicine should be procured from him which will restore the deranged system to its normal healthy condition.

243. Other Deceptions. Nearly all patent medicines contain some alcohol, and in many, the quantity of alcohol is far in excess of that found in the strongest wines. Tonics and bitters advertised as a cure for spring fever and a worn-out system are scarcely more than cheap cocktails, as one writer has derisively called them, and the amount of alcohol in some widely advertised patent remedies is alarmingly large and almost equal to that of strong whisky.

[Illustration: FIG. 161.--Diagram showing the amount of alcohol in some alcoholic drinks and in one much used patent medicine.]

Some conscientious persons who would not touch beer, wine, whisky, or any other intoxicating drink consume patent remedies containing large quantities of alcohol and thus unintentionally expose themselves to mental and physical danger. In all cases of bodily disorder, the only safe course is to consult a physician who has devoted himself to the study of the body and the methods by which a disordered system may be restored to health.

CHAPTER XXVI

NITROGEN AND ITS RELATION TO PLANTS

244. Nitrogen. A substance which plays an important part in animal and plant life is nitrogen. Soil and the fertilizers which enrich it, the plants which grow on it, and the animals which feed on these, all contain nitrogen or nitrogenous compounds. The atmosphere, which we ordinarily think of as a storehouse of oxygen, contains far more nitrogen than oxygen, since four fifths of its whole weight is made up of this element.

Nitrogen is colorless, odorless, and tasteless. Air is composed chiefly of oxygen and nitrogen; if, therefore, the oxygen in a vessel filled with air can be made to unite with some other substance or can be removed, there will be a residue of nitrogen. This can be done by floating on water a light dish containing phosphorus, then igniting the phosphorus, and placing an inverted jar over the burning substance. The phosphorus in burning unites with the oxygen of the air and hence the gas that remains in the jar is chiefly nitrogen. It has the characteristics mentioned above and, in addition, does not combine readily with other substances.

245. Plant Food. Food is the course of energy in every living thing and is essential to both animal and plant life. Plants get their food from the lifeless matter which exists in the air and in the soil; while animals get their food from plants. It is true that man and many other animals eat fleshy foods and depend upon them for partial sustenance, but the ultimate source of all animal food is plant life, since meat-producing animals live upon plant growth.

CHAPTER XXVI

Plants get their food from the air, the soil, and moisture. From the air, the leaves take carbon dioxide and water and transform them into starchy food; from the soil, the roots take water rich in mineral matters dissolved from the soil. From the substances thus gathered, the plant lives and builds up its structure.

A food substance necessary to plant life and growth is nitrogen. Since a vast store of nitrogen exists in the air, it would seem that plants should never lack for this food, but most plants are unable to make use of the boundless store of atmospheric nitrogen, because they do not possess the power of abstracting nitrogen from the air. For this reason, they have to depend solely upon nitrogenous compounds which are present in the soil and are soluble in water. The soluble nitrogenous soil compounds are absorbed by roots and are utilized by plants for food.

246. The Poverty of the Soil. Plant roots are constantly taking nitrogen and its compounds from the soil. If crops which grow from the soil are removed year after year, the soil becomes poorer in nitrogen, and finally possesses too little of it to support vigorous and healthy plant life. The nitrogen of the soil can be restored if we add to it a fertilizer containing nitrogen compounds which are soluble in water. Decayed vegetable matter contains large quantities of nitrogen compounds, and hence if decayed vegetation is placed upon soil or is plowed into soil, it acts as a fertilizer, returning to the soil what was taken from it. Since man and all other animals subsist upon plants, their bodies likewise contain nitrogenous substances, and hence manure and waste animal matter is valuable as a fertilizer or soil restorer.

247. Bacteria as Nitrogen Gatherers. Soil from which crops are removed year after year usually becomes less fertile, but the soil from which crops of clover, peas, beans, or alfalfa have been removed is richer in nitrogen rather than poorer. This is because the roots of these plants often have on them tiny swellings, or tubercles, in which millions of certain bacteria live and multiply. These bacteria have the remarkable power of taking free nitrogen from the air in the soil and of combining it with other substances to form compounds which plants can use. The bacteria-made compounds dissolve in the soil water and are absorbed into the plant by the roots. So much nitrogen-containing

CHAPTER XXVI

material is made by the root bacteria of plants of the pea family that the soil in which they grow becomes somewhat richer in nitrogen, and if plants which cannot make nitrogen are subsequently planted in such a soil, they find there a store of nitrogen. A crop of peas, beans, or clover is equivalent to nitrogenous fertilizer and helps to make ready the soil for other crops.

[Illustration: FIG. 162.--Roots of soy bean having tubercle-bearing bacteria.]

248. Artificial Fertilizers. Plants need other foods besides nitrogen, and they exhaust the soil not only of nitrogen, but also of phosphorus and potash, since large quantities of these are necessary for plant life. There are many other substances absorbed from the soil by the plant, namely, iron, sodium, calcium, magnesium, but these are used in smaller quantities and the supply in the soil does not readily become exhausted.

Commercial fertilizers generally contain nitrogen, phosphorus, and potash in amounts varying with the requirements of the soil. Wheat requires a large amount of phosphorus and quickly exhausts the ground of that food stuff; a field which has supported a crop of wheat is particularly poor in phosphorus, and a satisfactory fertilizer for that land would necessarily contain a large percentage of phosphorus. The fertilizer to be used in a soil depends upon the character of the soil and upon the crops previously grown on it.

[Illustration: FIG. 163.--Water cultures of buckwheat: 1, with all the food elements; 2, without potash; 3, without nitrates.]

The quantity of fertilizer needed by the farmers of the world is enormous, and the problem of securing the necessary substances in quantities sufficient to satisfy the demand bids fair to be serious. But modern chemistry is at work on the problem, and already it is possible to make some nitrogen compounds on a commercial scale. When nitrogen gas is in contact with heated calcium carbide, a reaction takes place which results in the formation of calcium nitride, a compound suitable for enriching the soil. There are other commercial methods for obtaining nitrogen compounds which are suitable for absorption by plant roots.

Phosphorus is obtained from bone ash and from phosphate rock which is widely distributed over the surface of the earth. Bone ash and thousands of tons of phosphate rock are treated with sulphuric acid to form a phosphorus compound which is soluble in soil water and which, when added to soil, will be usable by the plants growing there.

The other important ingredient of most fertilizers is potash. Wood ashes are rich in potash and are a valuable addition to the soil. But the amount of potash thus obtained is far too limited to supply the needs of agriculture; and to-day the main sources of potash are the vast deposits of potassium salts found in Prussia.

Although Germany now furnishes the American farmer with the bulk of his potash, she may not do so much longer. In 1911 an indirect potash tax was levied by Germany on her best customer, the United States, to whom 15 million dollars' worth of potash had been sold the preceding year. This led Americans to inquire whether potash could not be obtained at home.

Geologists say that long ages ago Germany was submerged, that the waters slowly evaporated and that the various substances in the sea water were deposited in thick layers. The deposits thus left by the evaporation of the sea water gradually became hidden by sediment and soil, and lost to sight. From such deposits, potash is obtained. Geologists tell us that our own Western States were once submerged, and that the waters evaporated and disappeared from our land very much as they did from Germany. The Great Salt Lake of Utah is a relic of a great body of water. If it be true that waters once covered our Western States, there may be buried deposits of potash there, and to-day the search for the hidden treasure is going on with the energy and enthusiasm characteristic of America.

Another probable source of potash is seaweed. The sea is a vast reservoir of potash, and seaweed, especially the giant kelp, absorbs large quantities of this potash. A ton of dried kelp (dried by sun and wind) contains about 500 pounds of pure potash. The kelps are abundant, covering thousands of square miles in the Pacific Ocean, from Mexico to the Arctic Ocean.

CHAPTER XXVII

SOUND

249. The Senses. All the information which we possess of the world around us comes to us through the use of the senses of sight, hearing, taste, touch, and smell. Of the five senses, sight and hearing are generally considered the most valuable. In preceding Chapters we studied the important facts relative to light and the power of vision; it remains for us to study Sound as we studied Light, and to learn what we can of sound and the power to hear.

250. How Sound is Produced. If one investigates the source of any sound, he will always find that it is due to motion of some kind. A sudden noise is traced to the fall of an object, or to an explosion, or to a collision; in fact, is due to the motion of matter. A piano gives out sound whenever a player strikes the keys and sets in motion the various wires within the piano; speech and song are caused by the motion of chest, vocal cords, and lips.

[Illustration: FIG. 164.--Sprays of water show that the fork is in motion.]

If a large dinner bell is rung, its motion or vibration may be felt on touching it with the finger. If a tuning fork is made to give forth sound by striking it against the knee, or hitting it with a rubber hammer, and is then touched to the surface of water, small sprays of water will be thrown out, showing that the prongs of the fork are in rapid motion. (A rubber hammer is made by putting a piece of glass tubing through a rubber cork.)

If a light cork ball on the end of a thread is brought in contact with a sounding fork, the ball does not remain at rest, but vibrates back and forth, being driven by the moving prongs.

[Illustration: FIG. 165.--The ball does not remain at rest]

CHAPTER XXVII 193

These simple facts lead us to conclude that all sound is due to the motion of matter, and that a sounding body of any kind is in rapid motion.

251. **Sound is carried by Matter.** In most cases sound reaches the ear through the air; but air is not the only medium through which sound is carried. A loud noise will startle fish, and cause them to dart away, so we conclude that the sound must have reached them through the water. An Indian puts his ear to the ground in order to detect distant footsteps, because sounds too faint to be heard through the air are comparatively clear when transmitted through the earth. A gentle tapping at one end of a long table can be distinctly heard at the opposite end if the ear is pressed against the table; if the ear is removed from the wood, the sound of tapping is much fainter, showing that wood transmits sound more readily than air. We see therefore that sound can be transmitted to the ear by solids, liquids, or gases.

Matter of any kind can transmit sound to the ear. The following experiments will show that matter is necessary for transmission. Attach a small toy bell to a glass rod (Fig. 166) by means of a rubber tube and pass the rod through one of two openings in a rubber cork. Insert the cork in a strong flask containing a small quantity of water and shake the bell, noting the sound produced. Then heat the flask, allowing the water to boil briskly, and after the boiling has continued for a few minutes remove the flame and instantly close up the second opening by inserting a glass stopper. Now shake the flask and note that the sound is very much fainter than at first. As the flask was warmed, air was rapidly expelled; so that when the flask was shaken the second time, less air was present to transmit the sound. If the glass stopper is removed and the air is allowed to reenter the flask, the loudness of the sound immediately increases.

[Illustration: FIG. 166.--Sound is carried by the air.]

Since the sound of the bell grows fainter as air is removed, we infer that there would be no sound if all the air were removed from the flask; that is to say, sound cannot be transmitted through empty space or a vacuum. If sound is to reach our ears, it must be through the agency of matter, such as wood, water, or air, etc.

CHAPTER XXVII

252. How Sound is transmitted through Air. We saw in Section 250 that sound can always be traced to the motion or vibration of matter. It is impossible to conceive of an object being set into sudden and continued motion without disturbing the air immediately surrounding it. A sounding body always disturbs and throws into vibration the air around it, and the air particles which receive motion from a sounding body transmit their motion to neighboring particles, these in turn to the next adjacent particles, and so on until the motion has traveled to very great distances. The manner in which vibratory motion is transmitted by the atmosphere must be unusual in character, since no motion of the air is apparent, and since in the stillness of night when "not a breath of air" is stirring, the shriek of a railroad whistle miles distant may be heard with perfect clearness. Moreover, the most delicate notes of a violin can be heard in the remotest corners of a concert hall, when not the slightest motion of the air can be seen or felt.

In our study of the atmosphere we saw that air can be compressed and rarefied; in other words, we saw that air is very elastic. It can be shown experimentally that whenever an elastic body in motion comes in contact with a body at rest, the moving body transfers its motion to the second body and then comes to rest itself. Let two billiard balls be suspended in the manner indicated in Figure 167. If one of the balls is drawn aside and is then allowed to fall against the other, the second ball is driven outward to practically the height from which the first ball fell and the first ball comes to rest.

[Illustration: FIG. 167.--Elastic balls.]

[Illustration: FIG. 168.--Suspended billiard balls.]

If a number of balls are arranged in line as in Figure 168 or Figure 169, and the end ball is raised and then allowed to fall, or if *A* is pushed against *C*, the last ball *B* will move outward alone, with a force nearly equal to that originally possessed by *A* and to a distance nearly equal to that through which *A* moved. But there will be no *visible* motion of the intervening balls. The force of the moving ball *A* is given to the second ball, and the second ball in turn gives the motion to the third, and so on throughout the entire number, until *B* is reached. But *B* has no ball to give its motion to, hence *B* itself moves outward, and moves

CHAPTER XXVII

with a force nearly equal to that originally imparted by *A* and to a distance nearly equal to that through which *A* fell. Motion at *A* is transmitted to *B* without any perceptible motion of the balls lying between these points. Similarly the particles of air set into motion by a sounding body impart their motion to each other, the motion being transmitted onward without any perceptible motion of the air itself. When this motion reaches the ear, it sets the drum of the ear into vibration, and these vibrations are in turn transmitted to the auditory nerves, which interpret the motion as sound.

[Illustration: FIG. 169.--Elastic balls transmit motion.]

[Illustration: FIG. 170.--When a ball meets more than one ball, it divides its motion.]

253. Why Sound dies away with Distance. Since the last ball *B* is driven outward with a force nearly equal to that possessed by *A*, it would seem that the effect on the ear drum should be independent of distance and that a sound should be heard as distinctly when remote as when near. But we know from experience that this is not true, because the more distant the source of sound, the fainter the impression; and finally, if the distance between the source of sound and the hearer becomes too great, the sound disappears entirely and nothing is heard. The explanation of this well-known fact is found in a further study of the elastic balls (Fig. 170). If *A* hits two balls instead of one, the energy possessed by *A* is given in part to one ball, and in part to the other, so that neither obtains the full amount. These balls, having each received less than the original energy, have less to transmit; each of these balls in turn meets with others, and hence the motion becomes more and more distributed, and distant balls receive less and less impetus. The energy finally given becomes too slight to affect neighboring balls, and the system comes to rest. This is what occurs in the atmosphere; a moving air particle meets not one but many adjacent air particles, and each of these receives a portion of the original energy and transmits a portion. When the original disturbance becomes scattered over a large number of air particles, the energy given to any one air particle becomes correspondingly small, and finally the energy becomes so small that further particles are not affected; beyond this limit the sound cannot be heard.

CHAPTER XXVII

If an air particle transmitted motion only to those air particles directly in line with it, we should not be able to detect sound unless the ear were in direct line with the source. The fact that an air particle divides its motion among all particles which it touches, that is, among those on the sides as well as those in front, makes it possible to hear sound in all directions. A good speaker is heard not only by those directly in front of him, but by those on the side, and even behind him.

254. Velocity of Sound. The transmission of motion from particle to particle does not occur instantaneously, but requires time. If the distance is short, so that few air particles are involved, the time required for transmission is very brief, and the sound is heard at practically the instant it is made. Ordinarily we are not conscious that it requires time for sound to travel from its source to our ears, because the distance involved is too short. At other times we recognize that there is a delay; for example, thunder reaches our ears after the lightning which caused the thunder has completely disappeared. If the storm is near, the interval of time between the lightning and the thunder is brief, because the sound does not have far to travel; if the storm is distant, the interval is much longer, corresponding to the greater distance through which the sound travels. Sound does not move instantaneously, but requires time for its transmission. The report of a distant cannon is heard after the flash and smoke are seen; the report of a near cannon is heard the instant the flash is seen.

The speed with which sounds travels through the air, or its velocity, was first measured by noting the interval (54.6 seconds) which elapsed between the flash of a cannon and the sound of the report. The distance of the cannon from the observer was measured and found to be 61,045 feet, and by dividing this distance by the number of seconds, we find that the distance traveled by sound in one second is approximately 1118 feet.

High notes and low notes, soft notes and shrill notes, all travel at the same rate. If bass notes traveled faster or slower than soprano notes, or if the delicate tones of the violin traveled faster or slower than the tones of a drum, music would be practically impossible, because at a distance from the source of sound the various tones which should be in unison would be out of time--some arriving late, some early.

CHAPTER XXVII 197

255. **Sound Waves.** Practically everyone knows that a hammock hung with long ropes swings or vibrates more slowly than one hung with short ropes, and that a stone suspended by a long string swings more slowly than one suspended by a short string. No two rocking chairs vibrate in the same way unless they are exactly alike in shape, size, and material. An object when disturbed vibrates in a manner peculiar to itself, the vibration being slow, as in the case of the long-roped swing, or quick, as in the case of the short-roped swing. The time required for a single swing or vibration is called the *period* of the body, and everything that can vibrate has a characteristic period. Size and shape determine to a large degree the period of a body; for example, a short, thick tuning fork vibrates more rapidly than a tall slender fork.

[Illustration: FIG. 171.--The two hammocks swing differently.]

Some tuning forks when struck vibrate so rapidly that the prongs move back and forth more than 5000 times per second, while other tuning forks vibrate so slowly that the vibrations do not exceed 50 per second. In either case the distance through which the prongs move is very small and the period is very short, so that the eye can seldom detect the movement itself. That the prongs are in motion, however, is seen by the action of a pith ball when brought in contact with the prongs (see Section 250).

[Illustration: FIG. 172.--The pitch given out by a fork depends upon its shape.]

The disturbance created by a vibrating body is called a wave.

256. **Waves.** While the disturbance which travels out from a sounding body is commonly called a wave, it is by no means like the type of wave best known to us, namely, the water wave.

If a closely coiled heavy wire is suspended as in Figure 173 and the weight is drawn down and then released, the coil will assume the appearance shown; there is clearly an overcrowding or condensation in some places, and a spreading out or rarefaction in other places. The pulse of condensation and rarefaction which travels the length of the wire is called a wave, although it bears little or no resemblance to the familiar water wave. Sound waves are similar to the waves formed in the stretched coil.

[Illustration: FIG. 173.--Waves in a coiled wire.]

Sound waves may be said to consist of a series of condensations and rarefactions, and the distance between two consecutive condensations and rarefactions may be defined as the wave length.

257. How One Sounding Body produces Sound in Another Body. In Section 255 we saw that any object when disturbed vibrates in a manner peculiar to itself,--its natural period,--a long-roped hammock vibrating slowly and a short-roped hammock vibrating rapidly. From observation we learn that it requires but little force to cause a body to vibrate in its natural period. If a sounding body is near a body which has the same period as itself, the pulses of air produced by the sounding body will, although very small, set the second body into motion and cause it to make a faint sound. When a piano is being played, we are often startled to find that a window pane or an ornament responds to some note of the piano. If two tuning forks of exactly identical periods (that is, of the same frequency) are placed on a table as in Figure 174, and one is struck so as to give forth a clear sound, the second fork will likewise vibrate, even though the two forks may be separated by several feet of air. We can readily see that the second fork is in motion, although it has not been struck, because it will set in motion a pith ball suspended beside it; at first the pith ball does not move, then it moves slightly, and finally bounces rapidly back and forth. If the periods of the two forks are not identical, but differ in the slightest degree, the second fork will not respond to the first fork, no matter how long or how loud the sound of the first fork. If we suppose that the fork vibrates 256 times each second, then 256 gentle pulses of air are produced each second, and these, traveling outward through the air, reach the silent fork and tend to set it in motion. A single pulse of air could not move the solid, heavy prongs, but the accumulated action of 256 vibrations per second soon makes itself felt, and the second fork begins to vibrate, at first gently, then gradually stronger, and finally an audible tone is given forth.

CHAPTER XXVII

[Illustration: FIG. 174.--When the first fork vibrates, the second responds.]

The cumulative power of feeble forces acting frequently at definite intervals is seen in many ways in everyday life. A small boy can easily swing a much larger boy, provided he gives the swing a gentle push in the right direction every time it passes him. But he must be careful to push at the proper instant, since otherwise his effort does not count for much; if he pushes forward when the swing is moving backward, he really hinders the motion; if he waits until the swing has moved considerably forward, his push counts for little. He must push at the proper instant; that is, the way in which his hand moves in giving the push must correspond exactly with the way in which the swing would naturally vibrate. A very striking experiment can be made by suspending from the ceiling a heavy weight and striking this weight gently at regular, properly timed intervals with a small cork hammer. Soon the pendulum, or weight, will be set swinging.

[Illustration: FIG. 175.--The hollow wooden box reënforces the sound.]

258. **Borrowed Sound.** Picture frames and ornaments sometimes buzz and give forth faint murmurs when a piano or organ is played. The waves sent out by a sounding body fall upon all surrounding objects and by their repeated action tend to throw these bodies into vibration. If the period of any one of the objects corresponds with the period of the sounding body, the gentle but frequent impulses affect the object, which responds by emitting a sound. If, however, the periods do not correspond, the action of the sound waves is not sufficiently powerful to throw the object into vibration, and no sound is heard. Bodies which respond in this way are said to be sympathetic and the response produced is called *resonance*. Seashells when held to the ear seem to contain the roar of the sea; this is because the air within the shell is set into sympathetic vibrations by some external tone. If the seashell were held to the ear in an absolutely quiet room, no sound would be heard, because there would be no external forces to set into vibration the air within the shell.

Tuning forks do not produce strong tones unless mounted on hollow wooden boxes (Fig. 175), whose size and shape are so adjusted that resonance occurs and strengthens the sound. When a human being talks or sings, the air within the mouth cavity is thrown into sympathetic vibration and strengthens the otherwise feeble tone of the speaker.

259. Echo. If one shouts in a forest, the sound is sometimes heard a second time a second or two later. This is because sound is reflected when it strikes a large obstructing surface. If the sound waves resulting from the shout meet a cliff or a mountain, they are reflected back, and on reaching the ear produce a later sensation of sound.

By observation it has been found that the ear cannot distinguish sounds which are less than one tenth of a second apart; that is, if two sounds follow each other at an interval less than one tenth of a second, the ear recognizes not two sounds, but one. This explains why a speaker can be heard better indoors than in the open air. In the average building, the walls are so close that the reflected waves have but a short distance to travel, and hence reach the ear at practically the same time as those which come directly from the speaker. In the open, there are no reflecting walls or surfaces, and the original sound has no reënforcement from reflection.

If the reflected waves reach the ear too late to blend with the original sound, that is, come later than one tenth of a second after the first impression, an echo is heard. What we call the rolling of thunder is really the reflection and re-reflection of the original thunder from cloud and cliff.

Some halls are so large that the reflected sounds cause a confusion of echoes, but this difficulty can be lessened by hanging draperies, which break the reflection.

260. Motion does not always produce Sound. While we know that all sound can be traced to motion, we know equally well that motion does not always produce sound. The hammock swinging in the breeze does not give forth a sound; the flag floating in the air does not give forth a sound unless blown violently by the wind; a card moved slowly through the air does not produce sound, but if the card is moved rapidly back and forth, a sound becomes audible.

CHAPTER XXVII 201

Motion, in order to produce sound, must be rapid; a ball attached to a string and moved slowly through the air produces no sound, but the same ball, whirled rapidly, produces a distinct buzz, which becomes stronger and stronger the faster the ball is whirled.

261. **Noise and Music.** When the rapid motions which produce sound are irregular, we hear noise; when the motions are regular and definite, we have a musical tone; the rattling of carriage wheels on stones, the roar of waves, the rustling of leaves are noise, not music. In all these illustrations we have rapid but irregular motion; no two stones strike the wheel in exactly the same way, no two waves produce pulses in the air of exactly the same character, no two leaves rustle in precisely the same way. The disturbances which reach the ear from carriage, waves, and leaves are irregular both in time and strength, and irritate the ear, causing the sensation which we call noise.

The tuning fork is musical. Here we have rapid, regular motion; vibrations follow each other at perfectly definite intervals, and the air disturbance produced by one vibration is exactly like the disturbance produced by a later vibration. The sound waves which reach the ear are regular in time and kind and strength, and we call the sensation music.

To produce noise a body must vibrate in such a way as to give short, quick shocks to the air; to produce music a body must not only impart short, quick shocks to the air, but must impart these shocks with unerring regularity and strength. A flickering light irritates the eye; a flickering sound or noise irritates the ear; both are painful because of the sudden and abrupt changes in effect which they cause, the former on the eye, the latter on the ear.

The only thing essential for the production of a musical sound is that the waves which reach the ear shall be rapid and regular; it is immaterial how these waves are produced. If a toothed wheel is mounted and slowly rotated, and a stiff card is held against the teeth of the wheel, a distinct tap is heard every time the card strikes the wheel. But if the wheel is rotated rapidly, the ear ceases to hear the various taps and recognizes a deep continuous musical tone. The blending of the individual taps, occurring at regular intervals, has produced a

CHAPTER XXVII 202

sustained musical tone. A similar result is obtained if a card is drawn slowly and then rapidly over the teeth of a comb.

[Illustration: FIG. 176.--A rotating disk.]

That musical tones are due to a succession of regularly timed impulses is shown most clearly by means of a rotating disk on which are cut two sets of holes, the outer set equally spaced, and the inner set unequally spaced (Fig. 176).

If, while the disk is rotating rapidly, a tube is held over the outside row and air is blown through the tube, a sustained musical tone will be heard. If, however, the tube is held, during the rotation of the disk, over the inner row of unequally spaced holes, the musical tone disappears, and a series of noises take its place. In the first case, the separate puffs of air followed each other regularly and blended into one tone; in the second case, the separate puffs of air followed each other at uncertain and irregular intervals and the result was noise.

Sound possesses a musical quality only when the waves or pulses follow each other at absolutely regular intervals.

262. The Effect of the Rapidity of Motion on the Musical Tone Produced. If the disk is rotated so slowly that less than about 16 puffs are produced in one second, only separate puffs are heard, and a musical tone is lacking; if, on the other hand, the disk is rotated in such a way that 16 puffs or more are produced in one second, the separate puffs will blend together to produce a continuous musical note of very low pitch. If the speed of the disk is increased so that the puffs become more frequent, the pitch of the resulting note rises; and at very high speeds the notes produced become so shrill and piercing as to be disagreeable to the ear. If the speed of the disk is lessened, the pitch falls correspondingly; and if the speed again becomes so low that less than 16 puffs are formed per second, the sustained sound disappears and a series of intermittent noises is produced.

263. The Pitch of a Note. By means of an apparatus called the siren, it is possible to calculate the number of vibrations producing any given musical note, such, for example, as middle C on the piano. If air is

forced continuously against the disk as it rotates, a series of puffs will be heard (Fig. 177).

If the disk turns fast enough, the puffs blend into a musical sound, whose pitch rises higher and higher as the disk moves faster and faster, and produces more and more puffs each second.

The instrument is so constructed that clockwork at the top registers the number of revolutions made by the disk in one second. The number of holes in the disk multiplied by the number of revolutions a second gives the number of puffs of air produced in one second. If we wish to find the number of vibrations which correspond to middle C on the piano, we increase the speed of the disk until the note given forth by the siren agrees with middle C as sounded on the piano, as nearly as the ear can judge; we then calculate the number of puffs of air which took place each second at that particular speed of the disk. In this way we find that middle C is due to about 256 vibrations per second; that is, a piano string must vibrate 256 times per second in order for the resultant note to be of pitch middle C. In a similar manner we determine the following frequencies:--

|do |re |mi |fa |sol |la |si |do | |C |D |E |F |G |A |B |C' | |256 |288 |320 |341 |384 |427 |480 |512 |

[Illustration: FIG. 177.--A siren.]

The pitch of pianos, from the lowest bass note to the very highest treble, varies from 27 to about 3500 vibrations per second. No human voice, however, has so great a range of tone; the highest soprano notes of women correspond to but 1000 vibrations a second, and the deepest bass of men falls but to 80 vibrations a second.

While the human voice is limited in its production of sound,--rarely falling below 80 vibrations a second and rarely exceeding 1000 vibrations a second,--the ear is by no means limited to that range in hearing. The chirrup of a sparrow, the shrill sound of a cricket, and the piercing shrieks of a locomotive are due to far greater frequencies, the number of vibrations at times equaling 38,000 per second or more.

CHAPTER XXVII

264. The Musical Scale. When we talk, the pitch of the voice changes constantly and adds variety and beauty to conversation; a speaker whose tone, or pitch, remains too constant is monotonous and dull, no matter how brilliant his thoughts may be.

While the pitch of the voice changes constantly, the changes are normally gradual and slight, and the different tones merge into each other imperceptibly. In music, however, there is a well-defined interval between even consecutive notes; for example, in the musical scale, middle C (do) with 256 vibrations is followed by D (re) with 288 vibrations, and the interval between these notes is sharp and well marked, even to an untrained ear. The interval between two notes is defined as the ratio of the frequencies; hence, the interval between C and D (do and re) is 288/256, or 9/8. Referring to Section 263, we see that the interval between C and E is 320/256, or 5/4, and the interval between C and C' is 512/256, or 2; the interval between any note and its octave is 2.

The successive notes in one octave of the musical scale are related as follows:--

Key of C	C	D	E	F	G	A	B	C'
No. of vibrations per sec.	256	288	320	341	384	427	480	512
Interval	9/8	5/4	4/3	3/2	5/3	15/8	2	

The intervals of F and A are not strictly 4/3 and 5/3, but are nearly so; if F made 341.3 vibrations per second instead of 341; and if A made 426.6 instead of 427, then the intervals would be exactly 4/3 and 5/3. Since the real difference is so slight, we can assume the simpler ratios without appreciable error.

Any eight notes whose frequencies are in the ratio of 9/8, 5/4, etc., will when played in succession give the familiar musical scale; for example, the deepest bass voice starts a musical scale whose notes have the frequencies 80, 90, 100, 107, 120, 133, 150, 160, but the intervals here are identical with those of a higher scale; the interval between C and D, 80 and 90, is 9/8, just as it was before when the frequencies were much greater; that is, 256 and 288. In singing "Home, Sweet Home," for example, a bass voice may start with a note vibrating only 132 times a second; while a tenor may start at a higher pitch, with a note vibrating 198 times per second, and a soprano would probably take a much higher range still, with an initial frequency of 528 vibrations per second. But no matter where the voices start, the intervals are always identical. The air as sung by the bass voice would be represented by *A*. The air as sung by the tenor voice would be represented by *B*. The air as sung by the soprano voice would be represented by *C*.

[Illustration: FIG. 178.--A song as sung by three voices of different pitch.]

CHAPTER XXVIII

MUSICAL INSTRUMENTS

265. Musical instruments maybe divided into three groups according to the different ways in which their tones are produced:--

First. The stringed instruments in which sound is produced by the vibration of stretched strings, as in the piano, violin, guitar, mandolin.

Second. The wind instruments in which sound is produced by the vibrations of definite columns of air, as in the organ, flute, cornet, trombone.

Third. The percussion instruments, in which sound is produced by the motion of stretched membranes, as in the drum, or by the motion of metal disks, as in the tambourines and cymbals.

266. Stringed Instruments. If the lid of a piano is opened, numerous wires are seen within; some long, some short, some coarse, some fine. Beneath each wire is a small felt hammer connected with the keys in such a way that when a key is pressed, a string is struck by a hammer and is thrown into vibration, thereby producing a tone.

CHAPTER XXVIII

206

If we press the lowest key, that is, the key giving forth the lowest pitch, we see that the longest wire is struck and set into vibration; if we press the highest key, that is, the key giving the highest pitch, we see that the shortest wire is struck. In addition, it is seen that the short wires which produce the high tones are fine, while the long wires which produce the low tones are coarse. The shorter and finer the wire, the higher the pitch of the tone produced. The longer and coarser the wire, the lower the pitch of the tone produced.

[Illustration: FIG. 179.--Piano wires seen from the back.]

The constant striking of the hammers against the strings stretches and loosens them and alters their pitch; for this reason each string is fastened to a screw which can be turned so as to tighten the string or to loosen it if necessary. The tuning of the piano is the adjustment of the strings so that each shall produce a tone of the right pitch. When the strings are tightened, the pitch rises; when the strings are loosened, the pitch falls.

What has been said of the piano applies as well to the violin, guitar, and mandolin. In the latter instruments the strings are few in number, generally four, as against eighty-eight in the piano; the hammer of the piano is replaced in the violin by the bow, and in the guitar by the fingers; varying pitches on any one string are obtained by sliding a finger of the left hand along the wire, and thus altering its length.

Frequent tuning is necessary, because the fine adjustments are easily disturbed. The piano is the best protected of all the stringed instruments, being inclosed by a heavy framework, even when in use.

[Illustration: FIG. 180.--Front view of an open piano.]

267. Strings and their Tones. Fasten a violin string to a wooden frame or box, as shown in Figure 181, stretching it by means of some convenient weight; then lay a yardstick along the box in order that the lengths may be determined accurately. If the stretched string is plucked with the fingers or bowed with the violin bow, a clear musical sound of definite pitch will be produced. Now divide the string into two equal parts by inserting the bridge midway between the two ends; and

CHAPTER XXVIII 207

pluck either half as before. The note given forth is of a decidedly higher pitch, and if by means of the siren we compare the pitches in the two cases, we find that the note sounded by the half wire is the octave of the note sounded by the entire wire; the frequency has been doubled by halving the length. If now the bridge is placed so that the string is divided into two unequal portions such as 1:3 and 2:3, and the shorter portion is plucked, the pitch will be still higher; the shorter the length plucked, the higher the pitch produced. This movable bridge corresponds to the finger of the violinist; the finger slides back and forth along the string, thus changing the length of the bowed portion and producing variations in pitch.

[Illustration: FIG. 181.--The length of a string influences the pitch.]

[Illustration: FIG. 182.--Only one half of the string is bowed, but both halves vibrate.]

If there were but one string, only one pitch could be sounded at any one time; the additional strings of the violin allow of the simultaneous production of several tones.

268. The Freedom of a String. Some stringed instruments give forth tones which are clear and sweet, but withal thin and lacking in richness and fullness. The tones sounded by two different strings may agree in pitch and loudness and yet produce quite different effects on the ear, because in one case the tone may be much more pleasing than in the other. The explanation of this is, that a string may vibrate in a number of different ways.

Touch the middle of a wire with the finger or a pencil (Fig. 182), thus separating it into two portions and draw a violin bow across the center of either half. Only one half of the entire string is struck, but the motion of this half is imparted to the other half and throws it into similar motion, and if a tiny A-shaped piece of paper or rider is placed upon the unbowed half, it is hurled off.

[Illustration: FIG. 183.--The string vibrates in three portions.]

CHAPTER XXVIII

If the wire is touched at a distance of one third its length and a bow is drawn across the middle of the smaller portion, the string will vibrate in three parts; we cannot always see these various motions in different parts of the string, but we know of their existence through the action of the riders.

Similarly, touching the wire one fourth of its length from an end makes it vibrate in four segments; touching it one fifth of its length makes it vibrate in five segments.

In the first case, the string vibrated as a whole string and also as two strings of half the length; hence, three tones must have been given out, one tone due to the entire string and two tones due to the segments. But we saw in Section 267 that halving the length of a string doubles the pitch of the resulting tone, and produces the octave of the original tone; hence a string vibrating as in Figure 183 gives forth three tones, one of which is the fundamental tone of the string, and two of which are the octave of the fundamental tone. Hence, the vibrating string produces two sensations, that of the fundamental note and of its octave.

[Illustration: FIG. 184.--When a string vibrates as a whole, it gives out the fundamental note.]

When a string is plucked in the middle without being held, it vibrates simply as a whole (Fig. 184), and gives forth but one note; this is called the fundamental. If the string is made to vibrate in two parts, it gives forth two notes, the fundamental, and a note one octave higher than the fundamental; this is called the first overtone. When the string is made to move as in Figure 183, three distinct motions are called forth, the motion of the entire string, the motion of the portion plucked, and the motion of the remaining unplucked portion of the string. Here, naturally, different tones arise, corresponding to the different modes of vibration. The note produced by the vibration of one third of the original string is called the second overtone.

The above experiments show that a string is able to vibrate in a number of different ways at the same time, and to emit simultaneously a number of different tones; also that the resulting complex sound consists of the fundamental and one or more overtones, and that the number of overtones present depends upon how and where the string is plucked.

[Illustration: FIG. 185.--A string can vibrate in a number of different ways simultaneously, and can produce different notes simultaneously.]

269. The Value of Overtones. The presence of overtones determines the quality of the sound produced. If the string vibrates as a whole merely, the tone given out is simple, and seems dull and characterless. If, on the other hand, it vibrates in such a way that overtones are present, the tone given forth is full and rich and the sensation is pleasing. A tuning fork cannot vibrate in more than one way, and hence has no overtones, and its tone, while clear and sweet, is far less pleasing than the same note produced by a violin or piano. The untrained ear is not conscious of overtones and recognizes only the strong dominant fundamental. The overtones blend in with the fundamental and are so inconspicuously present that we do not realize their existence; it is only when they are absent that we become aware of the beauty which they add to the music. A song played on tuning forks instead of on strings would be lifeless and unsatisfying because of the absence of overtones.

It is not necessary to hold finger or pencil at the points 1:3, 1:4, etc., in order to cause the string to vibrate in various ways; if a string is merely plucked or bowed at those places, the result will be the same. It is important to remember that no matter where a string of definite length is bowed, the note most distinctly heard will be the fundamental; but the quality of the emitted tone will vary with the bowing. For example, if a string is bowed in the middle, the effect will be far less pleasing than though it were bowed near the end. In the piano, the hammers are arranged so as to strike near one end of the string, at a distance of about 1:7 to 1:9; and hence a large number of overtones combine to reënforce and enrich the fundamental tone.

270. The Individuality of Instruments. It has been shown that a piano string when struck by a hammer, or a violin string when bowed, or a mandolin string when plucked, vibrates not only as a whole, but also in segments, and as a result gives forth not a simple tone, as we are

accustomed to think, but a very complex tone consisting of the fundamental and one or more overtones. If the string whose fundamental note is lower C (128 vibrations per second) is thrown into vibration, the tone produced may contain, in addition to the prominent fundamental, any one or more of the following overtones: C', G", C", E", C"', etc.

The number of overtones actually present depends upon a variety of circumstances: in the piano, it depends largely upon the location of the hammer; in the violin, upon the place and manner of bowing. Mechanical differences in construction account for prominent and numerous overtones in some instruments and for feeble and few overtones in others. The oboe, for example, is so constructed that only the high overtones are present, and hence the sound gives a "pungent" effect; the clarinet is so constructed that the even-numbered overtones are killed, and the presence of only odd-numbered overtones gives individuality to the instrument. In these two instruments we have vibrating air columns instead of vibrating strings, but the laws which govern vibrating strings are applicable to vibrating columns of air, as we shall see later. It is really the presence or absence of overtones which enables us to distinguish the note of the piano from that of the violin, flute, or clarinet. If overtones could be eliminated, then middle C, or any other note on the piano, would be indistinguishable from that same note sounded on any other instrument. The fundamental note in every instrument is the same, but the overtones vary with the instrument and lend individuality to each. The presence of high overtones in the oboe and the presence of odd-numbered overtones in the clarinet enable us to distinguish without fail the sounds given out by these instruments.

The richness and individuality of an instrument are due, not only to the overtones which accompany the fundamental, but also to the "forced" vibrations of the inclosing case, or of the sounding board. If a vibrating tuning fork is held in the hand, the sound will be inaudible except to those quite near; if, however, the base of the fork is held against the table, the sound is greatly intensified and becomes plainly audible throughout the room.

The vibrations of the fork are transmitted to the table top and throw it into vibrations similar to its own, and these additional vibrations intensify the original sound. Any fork, no matter what its frequency, can force the surface of the table into vibration, and hence the sound of any fork will be intensified by contact with a table or box.

This is equally true of strings; if stretched between two posts and bowed, the sound given out by a string is feeble, but if stretched over a sounding board, as in the piano, or over a wooden shell, as in the violin, the sound is intensified. Any note of the instrument will force the sounding body to vibrate, thus reënforcing the volume of sound, but some tones, or modes of vibration, do this more easily than others, and while the sounding board or shell always responds, it responds in varying degree. Here again we have not only enrichment of sound but also individuality of instruments.

271. The Kinds of Stringed Instruments. Stringed instruments may be grouped in the following three classes:--

a. Instruments in which the strings are set into motion by hammers--piano.

b. Instruments in which the strings are set into motion by bowing--violin, viola, violoncello, double bass.

c. Instruments in which the strings are set into motion by plucking--harp, guitar, mandolin.

[Illustration: FIG. 186.--1, violin; 2, viola; 3, violoncello; 4, double bass.]

a. The piano is too well known to need comment. In passing, it may be mentioned that in the construction of the modern concert piano approximately 40,000 separate pieces of material are used. The large number of pieces is due, partly, to the fact that the single string corresponding to any one key is usually replaced by no less than three or four similar strings in order that greater volume of sound may be obtained. The hammer connected to a key strikes four or more strings instead of one, and hence produces a greater volume of tone.

b. The viola is larger than the violin, has heavier and thicker strings, and is pitched to a lower key; in all other respects the two are similar. The violoncello, because of the length and thickness of its strings, is pitched a whole octave lower than the violin; otherwise it is similar. The unusual length and thickness of the strings of the double bass make it produce very low notes, so that it is ordinarily looked upon as the "bass voice" of the orchestra.

c. The harp has always been considered one of the most pleasing and perfect of musical instruments. Here the skilled performer has absolutely free scope for his genius, because his fingers can pluck the strings at will and hence regulate the overtones, and his feet can regulate at will the tension, and hence the pitch of the strings.

Guitar and mandolin are agreeable instruments for amateurs, but are never used in orchestral music.

[Illustration: FIG. 187.--A harp.]

272. Wind Instruments. In the so-called wind instruments, sound is produced by vibrating columns of air inclosed in tubes or pipes of different lengths. The air column is thrown into vibration either directly, by blowing across a narrow opening at one end of a pipe as in the case of the whistle, or indirectly, by exciting vibrations in a thin strip of wood or metal, called a reed, which in turn communicates its vibrations to the air column within.

The shorter the air column, the higher the pitch. This agrees with the law of vibrating strings which gives high pitches for short lengths.

[Illustration: FIG. 188.--Open organ pipes of different pitch.]

The pitch of the sound emitted by a column of air vibrating within a pipe varies according to the following laws:

1. The shorter the pipe, the higher the pitch.

2. The pitch of a note emitted by an open pipe is one octave higher than that of a closed pipe of equal length.

CHAPTER XXVIII 213

3. Air columns vibrate in segments just as do strings, and the tone emitted by a pipe of given length is complex, consisting of the fundamental and one or more overtones. The greater the number of overtones present, the richer the tone produced.

273. How the Various Pitches are Produced. With a pipe of fixed length, for example, the clarinet (Fig. 189, 1), different pitches are obtained by pressing keys which open holes in the tube and thus shorten or lengthen the vibrating air column and produce a rise or fall in pitch. Changes in pitch are also produced by variation in the player's breathing. By blowing hard or gently, the number of vibrations of the reed is increased or decreased and hence the pitch is altered.

[Illustration: FIG. 189--1, clarinet; 2, oboe; 3, flute.]

In the oboe (Fig. 189, 2) the vibrating air column is set into motion by means of two thin pieces of wood or metal placed in the mouthpiece of the tube. Variations in pitch are produced as in the clarinet by means of stops and varied breathing. In the flute, the air is set into motion by direct blowing from the mouth, as is done, for instance, when we blow into a bottle or key.

The sound given out by organ pipes is due to air blown across a sharp edge at the opening of a narrow tube. The air forced across the sharp edge is thrown into vibration and communicates its vibration to the air within the organ pipe. For different pitches, pipes of different lengths are used: for very low pitches long, closed pipes are used; for very high pitches short, open pipes are used. The mechanism of the organ is such that pressing a key allows the air to rush into the communicating pipe and a sound is produced characteristic of the length of the pipe.

[Illustration: FIG. 190.--1, horn; 2, trumpet; 3, trombone.]

[Illustration: FIG. 191.--1, kettledrum; 2, bass drum; 3, cymbals.]

[Illustration: FIG. 192.--The seating arrangement of the Philadelphia orchestra.]

CHAPTER XXIX 214

In the brass wind instruments such as horn, trombone, and trumpet, the lips of the player vibrate and excite the air within. Varying pitches are obtained partly by the varying wind pressure of the musician; if he breathes fast, the pitch rises; if he breathes slowly, the pitch falls. All of these instruments, however, except the trombone possess some valves which, on being pressed, vary the length of the tube and alter the pitch accordingly. In the trombone, valves are replaced by a section which slides in and out and shortens or lengthens the tube.

274. The Percussion Instruments. The percussion instruments, including kettledrums, bass drums, and cymbals, are the least important of all the musical instruments; and are usually of service merely in adding to the excitement and general effect of an orchestra.

In orchestral music the various instruments are grouped somewhat as shown in Figure 192.

CHAPTER XXIX

SPEAKING AND HEARING

[Illustration: FIG. 193.--The vibration of the vocal cords produces the sound of the human voice.]

275. Speech. The human voice is the most perfect of musical instruments. Within the throat, two elastic bands are attached to the windpipe at the place commonly called Adam's apple; these flexible bands have received the name of vocal cords, since by their vibration all speech is produced. In ordinary breathing, the cords are loose and are separated by a wide opening through which air enters and leaves the lungs. When we wish to speak, muscular effort stretches the cords, draws them closer together, and reduces the opening between them to a narrow slit, as in the case of the organ pipe. If air from the lungs is sent through the narrow slit, the vocal cords or bands are thrown into rapid vibration and produce sound. The pitch of the sound depends upon the tension of the stretched membranes, and since this can be altered by muscular action, the voice can be modulated at will. In times

of excitement, when the muscles of the body in general are in a state of great tension, the pitch is likely to be uncommonly high.

Women's voices are higher than men's because the vocal cords are shorter and finer; even though muscular tension is relaxed and the cords are made looser, the pitch of a woman's voice does not fall so low as that of a man's voice since his cords are naturally much longer and coarser. The difference between a soprano and an alto voice is merely one of length and tension of the vocal cords.

Successful singing is possible only when the vocal cords are readily flexible and when the singer can supply a steady, continuous blast of air through the slit between the cords. The hoarseness which frequently accompanies cold in the head is due to the thickening of the mucous membrane and to the filling up of the slit with mucus, because when this happens, the vocal cords cannot vibrate properly.

The sounds produced by the vocal cords are transformed into speech by the help of the tongue and lips, which modify the shape of the mouth cavity. Some of the lower animals have a speaking apparatus similar to our own, but they cannot perfectly transform sound into speech. The birds use their vocal cords to beautiful advantage in singing, far surpassing us in many ways, but the power of speech is lacking.

276. The Ear. The pulses created in the air by a sounding body are received by the ear and the impulses which they impart to the auditory nerve pass to the brain and we become conscious of a sound. The ear is capable of marvelous discrimination and accuracy. "In order to form an idea of the extent of this power imagine an auditor in a large music hall where a full band and chorus are performing. Here, there are sounds mingled together of all varieties of pitch, loudness, and quality; stringed instruments, wood instruments, brass instruments, and voices, of many different kinds. And in addition to these there may be all sorts of accidental and irregular sounds and noises, such as the trampling and shuffling of feet, the hum of voices, the rustle of dress, the creaking of doors, and many others. Now it must be remembered that the only means the ear has of becoming aware of these simultaneous sounds is by the condensations and rarefactions which

CHAPTER XXIX 216

reach it; and yet when the sound wave meets the nerves, the nerves single out each individual element, and convey to the mind of the hearer, not only the tones and notes of every instrument in the orchestra, but the character of every accidental noise; and almost as distinctly as if each single tone or noise were heard alone."--POLE.

[Illustration: FIG. 194.--The ear.]

277. The Structure of the Ear. The external portion of the ear acts as a funnel for catching sound waves and leading them into the canal, where they strike upon the ear drum, or tympanic membrane, and throw it into vibration. Unless the ear drum is very flexible there cannot be perfect response to the sound waves which fall upon it; for this reason, the glands of the canal secrete a wax which moistens the membrane and keeps it flexible. Lying directly back of the tympanic membrane is a cavity filled with air which enters by the Eustachian tube; from the throat air enters the Eustachian tube, moves along it, and passes into the ear cavity. The dull crackling noise noticed in the ear when one swallows is due to the entrance and exit of air in the tube. Several small bones stretch across the upper portion of the cavity and make a bridge, so to speak, from the ear drum to the far wall of the cavity. It is by means of these three bones that the vibrations of the ear drum are transmitted to the inner wall of the cavity. Behind the first cavity is a second cavity so complex and irregular that it is called the labyrinth of the ear. This labyrinth is filled with a fluid in which are spread out the delicate sensitive fibers of the auditory nerves; and it is to these that the vibrations must be transmitted.

Suppose a note of 800 vibrations per second is sung. Then 800 pulses of air will reach the ear each second, and the ear drum, being flexible, will respond and will vibrate at the same rate. The vibration of the ear drum will be transmitted by the three bones and the fluid to the fibers of the auditory nerves. The impulses imparted to the auditory nerve reach the brain and in some unknown way are translated into sound.

278. Care of the Ear. Most catarrhal troubles are accompanied by an oversupply of mucus which frequently clogs up the Eustachian tube and produces deafness. For the same reason, colds and sore throat

CHAPTER XXIX 217

sometimes induce temporary deafness.

The wax of the ear is essential for flexibility of the ear drum; if an extra amount accumulates, it can be got rid of by bathing the ear in hot water, since the heat will melt the wax. The wax should never be picked out with pin or sharp object except by a physician, lest injury be done to the tympanic membrane.

279. The Phonograph. The invention of the phonograph by Edison in 1878 marked a new era in the popularity and dissemination of music. Up to that time, household music was limited to those who were rich enough to possess a real musical instrument, and who in addition had the understanding and the skill to use the instrument. The invention of the phonograph has brought music to thousands of homes possessed of neither wealth nor skill. That the music reproduced by a phonograph is not always of the highest order does not, in the least, detract from the interest and wonder of the instrument. It can reproduce what it is called upon to reproduce, and if human nature demands the commonplace, the instrument will be made to satisfy the demand. On the other hand, speeches of famous men, national songs, magnificent opera selections, and other pleasing and instructive productions can be reproduced fairly accurately. In this way the phonograph, perhaps more than any other recent invention, can carry to the "shut-ins" a lively glimpse of the outside world and its doings.

[Illustration: FIG. 195.--A vibrating tuning fork traces a curved line on smoked glass.]

The phonograph consists of a cylinder or disk of wax upon which the vibrations of a sensitive diaphragm are recorded by means of a fine metal point. The action of the pointer in reporting the vibrations of a diaphragm is easily understood by reference to a tuning fork. Fasten a stiff bristle to a tuning fork by means of wax, allowing the end of the point to rest lightly upon a piece of smoked glass. If the glass is drawn under the bristle a straight line will be scratched on the glass, but if the tuning fork is struck so that the prongs vibrate back and forth, then the straight line changes to a wavy line and the type of wavy line depends upon the fork used.

In the phonograph, a diaphragm replaces the tuning fork and a cylinder (or a disk) coated with wax replaces the glass plate. When the speaker talks or the singer sings, his voice strikes against a delicate diaphragm and throws it into vibration, and the metal point attached to it traces on the wax of a moving cylinder a groove of varying shape and appearance called the "record." Every variation in the speaker's voice is repeated in the vibrations of the metal disk and hence in the minute motion of the pointer and in the consequent record on the cylinder. The record thus made can be placed in any other phonograph and if the metal pointer of this new phonograph is made to pass over the tracing, the process is reversed and the speaker's voice is reproduced. The sound given out in the this way is faint and weak, but can be strengthened by means of a trumpet attached to the phonograph.

[Illustration: FIG. 196.--A phonograph. In this machine the cylinder is replaced by a revolving disk.]

CHAPTER XXX

ELECTRICITY

280. Many animals possess the five senses, but only man possesses constructive, creative power, and is able to build on the information gained through the senses. It is the constructive, creative power which raises man above the level of the beast and enables him to devise and fashion wonderful inventions. Among the most important of his inventions are those which relate to electricity; inventions such as trolley car, elevator, automobile, electric light, the telephone, the telegraph. Bell, by his superior constructive ability, made possible the practical use of the telephone, and Marconi that of wireless telegraphy. To these inventions might be added many others which have increased the efficiency and production of the business world and have decreased the labor and strain of domestic life.

[Illustration: FIG. 197.--A simple electric cell.]

CHAPTER XXX

281. **Electricity as first Obtained by Man.** Until modern times the only electricity known to us was that of the lightning flash, which man could neither hinder nor make. But in the year 1800, electricity in the form of a weak current was obtained by Volta of Italy in a very simple way; and even now our various electric batteries and cells are but a modification of that used by Volta and called a voltaic cell. A strip of copper and a strip of zinc are placed in a glass containing dilute sulphuric acid, a solution composed of oxygen, hydrogen, sulphur, and water. As soon as the plates are immersed in the acid solution, minute bubbles of gas rise from the zinc strip and it begins to waste away slowly. The solution gradually dissolves the zinc and at the same time gives up some of the hydrogen which it contains; but it has little or no effect on the copper, since there is no visible change in the copper strip.

If, now, the strips are connected by means of metal wires, the zinc wastes away rapidly, numerous bubbles of hydrogen pass over to the copper strip and collect on it, and a current of electricity flows through the connecting wires. Evidently, the source of the current is the chemical action between the zinc and the liquid.

Mere inspection of the connecting wire will not enable us to detect that a current is flowing, but there are various ways in which the current makes itself evident. If the ends of the wires attached to the strips are brought in contact with each other and then separated, a faint spark passes, and if the ends are placed on the tongue, a twinge is felt.

282. **Experiments which grew out of the Voltaic Cell.** Since chemical action on the zinc is the source of the current, it would seem reasonable to expect a current if the cell consisted of two zinc plates instead of one zinc plate and one copper plate. But when the copper strip is replaced by a zinc strip so that the cell consists of two similar plates, no current flows between them. In this case, chemical action is expended in heat rather than in the production of electricity and the liquid becomes hot. But if carbon and zinc are used, a current is again produced, the zinc dissolving away as before, and bubbles collecting on the carbon plate. By experiment it has been found that many different metals may be employed in the construction of an electric cell; for example, current may be obtained from a cell made with a zinc plate and a platinum plate, or from a cell made with a lead plate and a

CHAPTER XXX 220

copper plate. Then, too, some other chemical, such as bichromate of potassium, or ammonium chloride, may be used instead of dilute sulphuric acid.

Almost any two different substances will, under proper conditions, give a current, but the strength of the current is in some cases so weak as to be worthless for practical use, such as telephoning, or ringing a door bell. What is wanted is a strong, steady current, and our choice of material is limited to the substances which will give this result. Zinc and lead can be used, but the current resulting is weak and feeble, and for general use zinc and carbon are the most satisfactory.

283. Electrical Terms. The plates or strips used in making an electric cell are called electrodes; the zinc is called the negative electrode (-), and the carbon the positive electrode (+); the current is considered to flow through the wire from the + to the-electrode. As a rule, each electrode has attached to it a binding post to which wires can be quickly fastened.

The power that causes the current is called the electromotive force, and the value of the electromotive force, generally written E.M.F., of a cell depends upon the materials used.

When the cell consists of copper, zinc, and dilute sulphuric acid, the electromotive force has a definite value which is always the same no matter what the size or shape of the cell. But the E.M.F. has a decidedly different value in a cell composed of iron, copper, and chromic acid. Each combination of material has its own specific electromotive force.

284. The Disadvantage of a Simple Cell. When the poles of a simple voltaic cell are connected by a wire, the current thus produced slowly diminishes in strength and, after a short time, becomes feeble. Examination of the cell shows that the copper plate is covered with hydrogen bubbles. If, however, these bubbles are completely brushed away by means of a rod or stick, the current strength increases, but as the bubbles again gather on the + electrode the current strength diminishes, and when the bubbles form a thick film on the copper plate, the current is too weak to be of any practical value. The film of

bubbles weakens the current because it practically substitutes a hydrogen plate for a copper plate, and we saw in Section 282 that a change in any one of the materials of which a cell is composed changes the current.

This weakening of the current can be reduced mechanically by brushing away the bubbles as soon as they are formed; or chemically, by surrounding the copper plate with a substance which will combine with the free hydrogen and prevent it from passing onward to the copper plate.

[Illustration: FIG 198. The gravity cell.]

In practically all cells, the chemical method is used in preference to the mechanical one. The numerous types of cells in daily use differ chiefly in the devices employed for preventing the formation of hydrogen bubbles, or for disposing of them when formed. One of the best-known cells in which weakening of the current is prevented by chemical means is the so-called gravity cell.

285. The Gravity Cell. A large, irregular copper electrode is placed in the bottom of a jar (Fig. 198), and completely covered with a saturated solution of copper sulphate. Then a large, irregular zinc electrode is suspended from the top of the jar, and is completely covered with dilute sulphuric acid which does not mix with the copper sulphate, but floats on the top of it like oil on water. The hydrogen formed by the chemical action of the dilute sulphuric acid on the zinc moves toward the copper electrode, as in the simple voltaic cell. It does not reach the electrode, however, because, when it comes in contact with the copper sulphate, it changes places with the copper there, setting it free, but itself entering into the solution. The copper freed from the copper sulphate solution travels to the copper electrode, and is deposited on it in a clean, bright layer. Instead of a deposit of hydrogen there is a deposit of copper, and falling off in current is prevented.

The gravity cell is cheap, easy to construct, and of constant strength, and is in almost universal use in telegraphic work. Practically all small railroad stations and local telegraph offices use these cells.

[Illustration: FIG. 199.--A dry cell.]

286. Dry Cells. The gravity cell, while cheap and effective, is inconvenient for general use, owing to the fact that it cannot be easily transported, and the *dry cell* has largely supplanted all others, because of the ease with which it can be taken from place to place. This cell consists of a zinc cup, within which is a carbon rod; the space between the cup and rod is packed with a moist paste containing certain chemicals. The moist paste takes the place of the liquids used in other cells.

[Illustration: FIG. 200.--A battery of three cells.]

287. A Battery of Cells. The electromotive force of one cell may not give a current strong enough to ring a door bell or to operate a telephone. But by using a number of cells, called a battery, the current may be increased to almost any desired strength. If three cells are arranged as in Figure 200, so that the copper of one cell is connected with the zinc of another cell, the electromotive force of the battery will be three times as great as the E.M.F. of a single cell. If four cells are arranged in the same way, the E.M.F. of the battery is four times as great as the E.M.F. of a single cell; when five cells are combined, the resulting E.M.F. is five times as great.

CHAPTER XXXI

SOME USES OF ELECTRICITY

288. Heat. Any one who handles electric wires knows that they are more or less heated by the currents which flow through them. If three cells are arranged as in Figure 200 and the connecting wire is coarse, the heating of the wire is scarcely noticeable; but if a shorter wire of the same kind is used, the heat produced is slightly greater; and if the coarse wire is replaced by a short, fine wire, the heating of the wire becomes very marked. We are accustomed to say that a wire offers resistance to the flow of a current; that is, whenever a current meets resistance, heat is produced in much the same way as when

CHAPTER XXXI

mechanical motion meets an obstacle and spends its energy in friction. The flow of electricity along a wire can be compared to the flow of water through pipes: a small pipe offers a greater resistance to the flow of water than a large pipe; less water can be forced through a small pipe than through a large pipe, but the friction of the water against the sides of the small pipe is much greater than in the large one.

So it is with the electric current. In fine wires the resistance to the current is large and the energy of the battery is expended in heat rather than in current. If the heat thus produced is very great, serious consequences may arise; for example, the contact of a hot wire with wall paper or dry beams may cause fire. Insurance companies demand that the wires used in wiring a building for electric lights be of a size suitable to the current to be carried, otherwise they will not take the risk of insurance. The greater the current to be carried, the coarser is the wire required for safety.

289. Electric Stoves. It is often desirable to utilize the electric current for the production of heat. For example, trolley cars are heated by coils of wire under the seats. The coils offer so much resistance to the passage of a strong current through them that they become heated and warm the cars.

[Illustration: FIG. 201.--An electric iron on a metal stand.]

Some modern houses are so built that electricity is received into them from the great plants where it is generated, and by merely turning a switch or inserting a plug, electricity is constantly available. In consequence, many practical applications of electricity are possible, among which are flatiron and toaster.

[Illustration: FIG. 202.--The fine wires are strongly heated by the current which flows through them.]

Within the flatiron (Fig. 201), is a mass of fine wire coiled as shown in Figure 202; as soon as the iron is connected with the house supply of electricity, current flows through the fine wire which thus becomes strongly heated and gives off heat to the iron. The iron, when once heated, retains an even temperature as long as the current flows, and

CHAPTER XXXI

224

the laundress is, in consequence, free from the disadvantages of a slowly cooling iron, and of frequent substitution of a warm iron for a cold one. Electric irons are particularly valuable in summer, because they eliminate the necessity for a strong fire, and spare the housewife intense heat. In addition, the user is not confined to the laundry, but is free to seek the coolest part of the house, the only requisite being an electrical connection.

[Illustration: FIG. 203.--Bread can be toasted by electricity.]

The toaster (Fig. 203) is another useful electrical device, since by means of it toast may be made on a dining table or at a bedside. The small electrical stove, shown in Figure 204, is similar in principle to the flatiron, but in it the heating coil is arranged as shown in Figure 205. To the physician electric stoves are valuable, since his instruments can be sterilized in water heated by the stove; and that without fuel or odor of gas.

A convenient device is seen in the heating pad (Fig. 206), a substitute for a hot water bag. Embedded in some soft thick substance are the insulated wires in which heat is to be developed, and over this is placed a covering of felt.

[Illustration: FIG. 204.--An electric stove.]

290. Electric Lights. The incandescent bulbs which illuminate our buildings consist of a fine, hairlike thread inclosed in a glass bulb from which the air has been removed. When an electric current is sent through the delicate filament, it meets a strong resistance. The heat developed in overcoming the resistance is so great that it makes the filament a glowing mass. The absence of air prevents the filament from burning, and it merely glows and radiates the light.

[Illustration: FIG. 205.--The heating element in the electric stove.]

291. Blasting. Until recently, dynamiting was attended with serious danger, owing to the fact that the person who applied the torch to the fuse could not make a safe retreat before the explosion. Now a fine wire is inserted in the fuse, and when everything is in readiness, the

CHAPTER XXXI 225

ends of the wire are attached to the poles of a distant battery and the heat developed in the wire ignites the fuse.

[Illustration: FIG. 206.--An electric pad serves the same purpose as a hot water bag.]

292. Welding of Metals. Metals are fused and welded by the use of the electric current. The metal pieces which are to be welded are pressed together and a powerful current is passed through their junction. So great is the heat developed that the metals melt and fuse, and on cooling show perfect union.

293. Chemical Effects. _The Plating of Gold, Silver, and Other Metals._ If strips of lead or rods of carbon are connected to the terminals of an electric cell, as in Figure 208, and are then dipped into a solution of copper sulphate, the strip in connection with the negative terminal of the cell soon becomes thinly plated with a coating of copper. If a solution of silver nitrate is used in place of the copper sulphate, the coating formed will be of silver instead of copper. So long as the current flows and there is any metal present in the solution, the coating continues to form on the negative electrode, and becomes thicker and thicker.

[Illustration: FIG. 207.--An incandescent electric bulb.]

The process by which metal is taken out of solution, as silver out of silver nitrate and copper out of copper sulphate, and is in turn deposited as a coating on another substance, is called electroplating. An electric current can separate a liquid into some of its various constituents and to deposit one of the metal constituents on the negative electrode.

[Illustration: FIG. 208.--Carbon rods in a solution of copper sulphate.]

Since copper is constantly taken out of the solution of copper sulphate for deposit upon the negative electrode, the amount of copper remaining in the solution steadily decreases, and finally there is none of it left for deposit. In order to overcome this, the positive electrode should be made of the same metal as that which is to be deposited.

CHAPTER XXXI

The positive metal electrode gradually dissolves and replaces the metal lost from the solution by deposit and electroplating can continue as long as any positive electrode remains.

[Illustration: FIG. 209.--Plating spoons by electricity.]

Practically all silver, gold, and nickel plating is done in this way; machine, bicycle, and motor attachments are not solid, but are of cheaper material electrically plated with nickel. When spoons are to be plated, they are hung in a bath of silver nitrate side by side with a thick slab of pure silver, as in Figure 209. The spoons are connected with the negative terminal of the battery, while the slab of pure silver is connected with the positive terminal of the same battery. The length of time that the current flows determines the thickness of the plating.

294. How Pure Metal is obtained from Ore. When ore is mined, it contains in addition to the desired metal many other substances. In order to separate out the desired metal, the ore is placed in some suitable acid bath, and is connected with the positive terminal of a battery, thus taking the place of the silver slab in the last Section. When current flows, any pure metal which is present is dissolved out of the ore and is deposited on a convenient negative electrode, while the impurities remain in the ore or drop as sediment to the bottom of the vessel. Metals separated from the ore by electricity are called electrolytic metals and are the purest obtainable.

295. Printing. The ability of the electric current to decompose a liquid and to deposit a metal constituent has practically revolutionized the process of printing. Formerly, type was arranged and retained in position until the required number of impressions had been made, the type meanwhile being unavailable for other uses. Moreover, the printing of a second edition necessitated practically as great labor as did the first edition, the type being necessarily set afresh. Now, however, the type is set up and a mold of it is taken in wax. This mold is coated with graphite to make it a conductor and is then suspended in a bath of copper sulphate, side by side with a slab of pure copper. Current is sent through the solution as described in Section 293, until a thin coating of copper has been deposited on the mold. The mold is then taken from the bath, and the wax is replaced by some metal

which gives strength and support to the thin copper plate. From this copper plate, which is an exact reproduction of the original type, many thousand copies can be printed. The plate can be preserved and used from time to time for later editions, and the original type can be put back into the cases and used again.

CHAPTER XXXII

MODERN ELECTRICAL INVENTIONS

296. **An Electric Current acts like a Magnet.** In order to understand the action of the electric bell, we must consider a third effect which an electric current can cause. Connect some cells as shown in Figure 200 and close the circuit through a stout heavy copper wire, dipping a portion of the wire into fine iron filings. A thick cluster of filings will adhere to the wire (Fig. 210), and will continue to cling to it so long as the current flows. If the current is broken, the filings fall from the wire, and only so long as the current flows through the wire does the wire have power to attract iron filings. An electric current makes a wire equivalent to a magnet, giving it the power to attract iron filings.

[Illustration: FIG. 210.--A wire carrying current attracts iron filings.]

[Illustration: FIG. 211.--A loosely wound coil of wire.]

Although such a straight current bearing wire attracts iron filings, its power of attraction is very small; but its magnetic strength can be increased by coiling as in Figure 211. Such an arrangement of wire is known as a helix or solenoid, and is capable of lifting or pulling larger and more numerous filings and even good-sized pieces of iron, such as tacks. Filings do not adhere to the sides of the helix, but they cling in clusters to the ends of the coil. This shows that the ends of the helix have magnetic power but not the sides.

If a soft iron nail (Fig. 212) or its equivalent is slipped within the coil, the lifting and attractive power of the coil is increased, and comparatively heavy weights can be lifted.

[Illustration: FIG. 212.--Coil and soft iron rod.]

A coil of wire traversed by an electric current and containing a core of soft iron has the power of attracting and moving heavy iron objects; that is, it acts like a magnet. Such an arrangement is called an electromagnet. As soon as the current ceases to flow, the electromagnet loses its magnetic power and becomes merely iron and wire without magnetic attraction.

If many cells are used, the strength of the electromagnet is increased, and if the coil is wound closely, as in Figure 213, instead of loosely, as in Figure 211, the magnetic strength is still further increased. The strength of any electromagnet depends upon the number of coils wound on the iron core and upon the strength of the current which is sent through the coils.

[Illustration: FIG. 213.--An electromagnet.]

[Illustration: FIG. 214.--A horseshoe electromagnet is powerful enough to support heavy weights.]

To increase the strength of the electromagnet still further, the so-called horseshoe shape is used (Fig. 214). In such an arrangement there is practically the strength of two separate electromagnets.

297. The Electric Bell. The ringing of the electric bell is due to the attractive power of an electromagnet. By the pushing of a button (Fig. 215) connection is made with a battery, and current flows through the wire wound on the iron spools, and further to the screw P which presses against the soft iron strip or armature S; and from S the current flows back to the battery. As soon as the current flows, the coils become magnetic and attract the soft iron armature, drawing it forward and causing the clapper to strike the bell. In this position, S no longer touches the screw P, and hence there is no complete path for the electricity, and the current ceases. But the attractive, magnetic power of the coils stops as soon as the current ceases; hence there is nothing to hold the armature down, and it flies back to its former position. In doing this, however, the armature makes contact at P through the spring, and the current flows once more; as a result the

CHAPTER XXXII

coils again become magnets, the armature is again drawn forward, and the clapper again strikes the bell. But immediately afterwards the armature springs backward and makes contact at P and the entire operation is repeated. So long as we press the button this process continues producing what sounds like a continuous jingle; in reality the clapper strikes the bell every time a current passes through the electromagnet.

[Illustration: FIG. 215.--The electric bell.]

298. The Push Button. The push button is an essential part of every electric bell, because without it the bell either would not ring at all, or would ring incessantly until the cell was exhausted. When the push button is free, as in Figure 216, the cell terminals are not connected in an unbroken path, and hence the current does not flow. When, however, the button is pressed, the current has a complete path, provided there is the proper connection at S. That is, the pressure on the push button permits current to flow to the bell. The flow of this current then depends solely upon the connection at S, which is alternately made and broken, and in this way produces sound.

[Illustration: FIG. 216.--Push button.]

The sign "Bell out of order" is usually due to the fact that the battery is either temporarily or permanently exhausted. In warm weather the liquid in the cell may dry up and cause stoppage of the current. If fresh liquid is poured into the vessel so that the chemical action of the acid on the zinc is renewed, the current again flows. Another explanation of an out-of-order bell is that the liquid may have eaten up all the zinc; if this is the case, the insertion of a fresh strip of zinc will remove the difficulty and the current will flow. If dry cells are used, there is no remedy except in the purchase of new cells.

299. How Electricity may be lost to Use. In the electric bell, we saw that an air gap at the push button stopped the flow of electricity. If we cut the wire connecting the poles of a battery, the current ceases because an air gap intervenes and electricity does not readily pass through air. Many substances besides air stop the flow of electricity. If a strip of glass, rubber, mica, or paraffin is introduced anywhere in a

circuit, the current ceases. If a metal is inserted in the gap, the current again flows. Substances which, like an air gap, interfere with the flow of electricity are called non-conductors, or, more commonly, insulators. Substances which, like the earth, the human body, and all other moist objects, conduct electricity are conductors. If the telephone and electric light wires in our houses were not insulated by a covering of thread, or cloth, or other non conducting material, the electricity would escape into surrounding objects instead of flowing through the wire and producing sound and light.

In our city streets, the overhead wires are supported on glass knobs or are closely wrapped, in order to prevent the escape of electricity through the poles to the ground. In order to have a steady, dependable current, the wire carrying the current must be insulated.

Lack of insulation means not only the loss of current for practical uses, but also serious consequences in the event of the crossing of current-bearing wires. If two wires properly insulated touch each other, the currents flow along their respective wires unaltered; if, however, two uninsulated wires touch, some of the electricity flows from one to the other. Heat is developed as a result of this transference, and the heat thus developed is sometimes so great that fire occurs. For this reason, wires are heavily insulated and extra protection is provided at points where numerous wires touch or cross.

Conductors and insulators are necessary to the efficient and economic flow of a current, the insulator preventing the escape of electricity and lessening the danger of fire, and the conductor carrying the current.

300. The Telegraph. Telegraphy is the process of transmitting messages from place to place by means of an electric current. The principle underlying the action of the telegraph is the principle upon which the electric bell operates; namely, that a piece of soft iron becomes a magnet while a current flows around it, but loses its magnetism as soon as the current ceases.

In the electric bell, the electromagnet, clapper, push button, and battery are relatively near,--usually all are located in the same building; while in the telegraph the current may travel miles before it reaches the electromagnet and produces motion of the armature.

[Illustration: FIG. 217.--Diagram of the electric telegraph.]

The fundamental connections of the telegraph are shown in Figure 217. If the key K is pressed down by an operator in Philadelphia, the current from the battery (only one cell is shown for simplicity) flows through the line to New York, passes through the electromagnet M, and thence back to Philadelphia. As long as the key K is pressed down, the coil M acts as a magnet and attracts and holds fast the armature A; but as soon as K is released, the current is broken, M loses its magnetism, and the armature is pulled back by the spring D. By a mechanical device, tape is drawn uniformly under the light marker P attached to the armature. If K is closed for but a short time, the armature is drawn down for but a short interval, and the marker registers a dot on the tape. If K is closed for a longer time, a short dash is made by the marker, and, in general, the length of time that K is closed determines the length of the marks recorded on the tape. The telegraphic alphabet consists of dots and dashes and their various combinations, and hence an interpretation of the dot and dash symbols recorded on the tape is all that is necessary for the receiving of a telegraphic message.

The Morse telegraphic code, consisting of dots, dashes, and spaces, is given in Figure 218.

[Illustration:

A .-	H	O . .	U ..-		B -...	I ..	P	V ...-		C .. .	J -.-.	Q ..-.	W .--
D -..	K -.-	R . ..	X .-..		E .	L ---	S ...	Y		F .-.	M - -	T -	Z .... .
G --.	N -.		

FIG. 218.--The Morse telegraphic code.]

The telegraph is now such a universal means of communication between distant points that one wonders how business was conducted before its invention in 1832 by S.F.B. Morse.

[Illustration: FIG. 219.--The sounder.]

CHAPTER XXXII

301. Improvements. *The Sounder.* Shortly after the invention of telegraphy, operators learned that they could read the message by the click of the marker against a metal rod which took the place of the tape. In practically all telegraph offices of the present day the old-fashioned tape is replaced by the sounder, shown in Figure 219. When current flows, a lever, L, is drawn down by the electromagnet and strikes against a solid metal piece with a click; when the current is broken, the lever springs upward, strikes another metal piece and makes a different click. It is clear that the working of the key which starts and stops the current in this line will be imitated by the motion and the resulting clicks of the sounder. By means of these varying clicks of the sounder, the operator interprets the message.

[Illustration: FIG. 220.--Diagram of a modern telegraph system.]

The Relay. When a telegraph line is very long, the resistance of the wire is great, and the current which passes through the electromagnet is correspondingly weak, so feeble indeed that the armature must be made very thin and light in order to be affected by the makes and breaks in the current. The clicks of an armature light enough to respond to the weak current of a long wire are too faint to be recognized by the ear, and hence in such long circuits some device must be introduced whereby the effect is increased. This is usually done by installing at each station a local battery and a very delicate and sensitive electromagnet called the *relay.* Under these conditions the current of the main line is not sent through the sounder, but through the relay which opens and closes a local battery in connection with the strong sounder. For example, the relay is so arranged that current from the main line runs through it exactly as it runs through M in Figure 217. When current is made, the relay attracts an armature, which thereby closes a circuit in a local battery and thus causes a click of the sounder. When the current in the main line is broken, the relay loses its magnetic attraction, its armature springs back, connection is broken in the local circuit, and the sounder responds by allowing its armature to spring back with a sharp sound.

302. The Earth an Important Part of a Telegraphic System. We learned in Section 299 that electricity could flow through many different substances, one of which was the earth. In all ordinary telegraph lines,

advantage is taken of this fact to utilize the earth as a conductor and to dispense with one wire. Originally two wires were used, as in Figure 217; then it was found that a railroad track could be substituted for one wire, and later that the earth itself served equally well for a return wire. The present arrangement is shown in Figure 220, where there is but one wire, the circuit being completed by the earth. No fact in electricity seems more marvelous than that the thousands of messages flashing along the wires overhead are likewise traveling through the ground beneath. If it were not for this use of the earth as an unfailing conductor, the network of overhead wires in our city streets would be even more complex than it now is.

303. Advances in Telegraphy. The mechanical improvements in telegraphy have been so rapid that at present a single operator can easily send or receive forty words a minute. He can telegraph more quickly than the average person can write; and with a combination of the latest improvements the speed can be enormously increased. Recently, 1500 words were flashed from New York to Boston over a single wire in one second.

In actual practice messages are not ordinarily sent long distances over a direct line, but are automatically transferred to new lines at definite points. For example, a message from New York to Chicago does not travel along an uninterrupted path, but is automatically transferred at some point, such as Lancaster, to a second line which carries it on to Pittsburgh, where it is again transferred to a third line which takes it farther on to its destination.

CHAPTER XXXIII

MAGNETS AND CURRENTS

304. In the twelfth century, there was introduced into Europe from China a simple instrument which changed journeying on the sea from uncertain wandering to a definite, safe voyage. This instrument was the compass (Fig. 221), and because of the property of the compass needle (a magnet) to point unerringly north and south, sailors were

CHAPTER XXXIII

able to determine directions on the sea and to steer for the desired point.

[Illustration: FIG. 221.--The compass.]

Since an electric current is practically equivalent to a magnet (Section 296), it becomes necessary to know the most important facts relative to magnets, facts simple in themselves but of far-reaching value and consequences in electricity. Without a knowledge of the magnetic characteristics of currents, the construction of the motor would have been impossible, and trolley cars, electric fans, motor boats, and other equally well-known electrical contrivances would be unknown.

305. **The Attractive Power of a Magnet.** The magnet best known to us all is the compass needle, but for convenience we will use a magnetic needle in the shape of a bar larger and stronger than that employed in the compass. If we lay such a magnet on a pile of iron filings, it will be found on lifting the magnet that the filings cling to the ends in tufts, but leave it almost bare in the center (Fig. 222). The points of attraction at the two ends are called the poles of the magnet.

[Illustration: FIG. 222.--A magnet.]

If a delicately made magnet is suspended as in Figure 223, and is allowed to swing freely, it will always assume a definite north and south position. The pole which points north when the needle is suspended is called the north pole and is marked *N*, while the pole which points south when the needle is suspended is called the south pole and is marked *S*.

A freely suspended magnet points nearly north and south.

A magnet has two main points of attraction called respectively the north and south poles.

[Illustration: FIG. 223.--The magnetic needle.]

306. **The Extent of Magnetic Attraction.** If a thin sheet of paper or cardboard is laid over a strong, bar-shaped magnet and iron filings are

then gently strewn on the paper, the filings clearly indicate the position of the magnet beneath, and if the cardboard is gently tapped, the filings arrange themselves as shown in Figure 224. If the paper is held some distance above the magnet, the influence on the filings is less definite, and finally, if the paper is held very far away, the filings do not respond at all, but lie on the cardboard as dropped.

The magnetic power of a magnet, while not confined to the magnet itself, does not extend indefinitely into the surrounding region; the influence is strong near the magnet, but at a distance becomes so weak as to be inappreciable. The region around a magnet through which its magnetic force is felt is called the field of force, or simply the magnetic field, and the definite lines in which the filings arrange themselves are called lines of force.

[Illustration: FIG. 224.--Iron filings scattered over a magnet arrange themselves in definite lines.]

The magnetic power of a magnet is not limited to the magnet, but extends to a considerable distance in all directions.

307. The Influence of Magnets upon Each Other. If while our suspended magnetic needle is at rest in its characteristic north-and-south direction another magnet is brought near, the suspended magnet is turned; that is, motion is produced (Fig. 225). If the north pole of the free magnet is brought toward the south pole of the suspended magnet, the latter moves in such a way that the two poles *N* and *S* are as close together as possible. If the north pole of the free magnet is brought toward the north pole of the suspended magnet, the latter moves in such a way that the two poles *N* and *N* are as far apart as possible. In every case that can be tested, it is found that a north pole repels a north pole, and a south pole repels a south pole; but that a north and a south pole always attract each other.

[Illustration: FIG. 225.--A south pole attracts a north pole.]

The main facts relative to magnets may be summed up as follows:--

CHAPTER XXXIII

a. A magnet points nearly north and south if it is allowed to swing freely.

b. A magnet contains two unlike poles, one of which persistently points north, and the other of which as persistently points south, if allowed to swing freely.

c. Poles of the same name repel each other; poles of unlike name attract each other.

d. A magnet possesses the power of attracting certain substances, like iron, and this power of attraction is not limited to the magnet itself but extends into the region around the magnet.

308. Magnetic Properties of an Electric Current. If a current-bearing wire is really equivalent in its magnetic powers to a magnet, it must possess all of the characteristics mentioned in the preceding Section. We saw in Section 296 that a coiled wire through which current was flowing would attract iron filings at the two ends of the helix. That a coil through which current flows possesses the characteristics *a*, *b*, *c*, and *d* of a magnet is shown as follows:--

a, b. If a helix marked at one end with a red string is arranged so that it is free to rotate and a strong current is sent through it, the helix will immediately turn and face about until it points north and south. If it is disturbed from this position, it will slowly swing back until it occupies its characteristic north and south position. The end to which the string is attached will persistently point either north or south. If the current is sent through the coil in the opposite direction, the two poles exchange positions and the helix turns until the new north pole points north.

[Illustration: FIG. 226.--A helix through which current flows always points north and south, if it is free to rotate.]

c. If a coil conducting a current is held near a suspended magnet, one end of the helix will be found to attract the north pole of the magnet, while the opposite end will be found to repel the north pole of the magnet. In fact, the helix will be found to behave in every way as a magnet, with a north pole at one end and a south pole at the other. If

CHAPTER XXXIII 237

the current is sent through the helix in the opposite direction, the north and south poles exchange places.

[Illustration: FIG. 227.--A wire through which current flows is surrounded by a field of magnetic force.]

If the number of turns in the helix is reduced until but a single loop remains, the result is the same; the single loop acts like a flat magnet, one side of the loop always facing northward and one southward, and one face attracting the north pole of the suspended magnet and one repelling it.

d. If a wire is passed through a card and a strong current is sent through the wire, iron filings will, when sprinkled upon the card, arrange themselves in definite directions (Fig. 227). A wire carrying a current is surrounded by a magnetic field of force.

A magnetic needle held under a current-bearing wire turns on its pivot and finally comes to rest at an angle with the current. The fact that the needle is deflected by the wire shows that the magnetic power of the wire extends into the surrounding medium.

The magnetic properties of current electricity were discovered by Oersted of Denmark less than a hundred years ago; but since that time practically all important electrical machinery has been based upon one or more of the magnetic properties of electricity. The motors which drive our electric fans, our mills, and our trolley cars owe their existence entirely to the magnetic action of current electricity.

[Illustration: FIG. 228.--The coil turns in such a way that its north pole is opposite the south pole of the magnet.]

309. The Principle of the Motor. If a close coil of wire is suspended between the poles of a strong horseshoe magnet, it will not assume any characteristic position but will remain wherever placed. If, however, a current is sent through the wire, the coil faces about and assumes a definite position. This is because a coil, carrying a current, is equivalent to a magnet with a north and south face; and, in accordance with the magnetic laws, tends to move until its north face

is opposite the south pole of the horseshoe magnet, and its south face opposite the north pole of the magnet. If, when the coil is at rest in this position, the current is reversed, so that the north pole of the coil becomes a south pole and the former south pole becomes a north pole, the result is that like poles of coil and magnet face each other. But since like poles repel each other, the coil will move, and will rotate until its new north pole is opposite to the south pole of the magnet and its new south pole is opposite the north pole. By sending a strong current through the coil, the helix is made to rotate through a half turn; by reversing the current when the coil is at the half turn, the helix is made to continue its rotation and to swing through a whole turn. If the current could be repeatedly reversed just as the helix completed its half turn, the motion could be prolonged; periodic current reversal would produce continuous rotation. This is the principle of the motor.

[Illustration: FIG. 229.--Principle of the motor.]

It is easy to see that long-continued rotation would be impossible in the arrangement of Figure 228, since the twisting of the suspending wire would interfere with free motion. If the motor is to be used for continuous motion, some device must be employed by means of which the helix is capable of continued rotation around its support.

In practice, the rotating coil of a motor is arranged as shown in Figure 229. Wires from the coil terminate on metal disks and are securely soldered there. The coil and disks are supported by the strong and well-insulated rod *R*, which rests upon braces, but which nevertheless rotates freely with disks and coil. The current flows to the coil through the thin metal strips called brushes, which rest lightly upon the disks.

When the current which enters at *B* flows through the wire, the coil rotates, tending to set itself so that its north face is opposite the south face of the magnet. If, when the helix has just reached this position, the current is reversed--entering at *B'* instead of *B*--the poles of the coil are exchanged; the rotation, therefore, does not cease, but continues for another half turn. Proper reversals of the current are accompanied by continuous motion, and since the disk and shaft rotate with the coil, there is continuous rotation.

If a wheel is attached to the rotating shaft, weights can be lifted, and if a belt is attached to the wheel, the motion of the rotating helix can be transferred to machinery for practical use.

The rotating coil is usually spoken of as the armature, and the large magnet as the field magnet.

310. Mechanical Reversal of the Current. *The Commutator.* It is not possible by hand to reverse the current with sufficient rapidity and precision to insure uninterrupted rotation; moreover, the physical exertion of such frequent reversals is considerable. Hence, some mechanical device for periodically reversing the current is necessary, if the motor is to be of commercial value.

[Illustration: FIG. 230.--The commutator.]

The mechanical reversal of the current is accomplished by the use of the commutator, which is a metal ring split into halves, well insulated from each other and from the shaft. To each half of this ring is attached one of the ends of the armature wire. The brushes which carry the current are set on opposite sides of the ring and do not rotate. As armature, commutator, and shaft rotate, the brushes connect first with one segment of the commutator and then with the other. Since the circuit is arranged so that the current always enters the commutator through the brush B, the flow of the current into the coil is always through the segment in contact with B; but the segment in contact with B changes at every half turn of the coil, and hence the direction of the current through the coil changes periodically. As a result the coil rotates continuously, and produces motion so long as current is supplied from without.

311. The Practical Motor. A motor constructed in accordance with Section 309 would be of little value in practical everyday affairs; its armature rotates too slowly and with too little force. If a motor is to be of real service, its armature must rotate with sufficient strength to impart motion to the wheels of trolley cars and mills, to drive electric fans, and to set into activity many other forms of machinery.

The strength of a motor may be increased by replacing the singly coiled armature by one closely wound on an iron core; in some armatures there are thousands of turns of wire. The presence of soft iron within the armature (Section 296) causes greater attraction between the armature and the outside magnet, and hence greater force of motion. The magnetic strength of the field magnet influences greatly the speed of the armature; the stronger the field magnet the greater the motion, so electricians make every effort to strengthen their field magnets. The strongest known magnets are electromagnets, which, as we have seen, are merely coils of wire wound on an iron core. For this reason, the field magnet is usually an electromagnet.

When very powerful motors are necessary, the field magnet is so arranged that it has four or more poles instead of two; the armature likewise consists of several portions, and even the commutator may be very complex. But no matter how complex these various parts may seem to be, the principle is always that stated in Section 309, and the parts are limited to field magnet, commutator, and armature.

[Illustration: FIG. 231.--A modern power plant.]

[Illustration: FIG. 232.--The electric street car.]

The motor is of value because by means of it motion, or mechanical energy, is obtained from an electric current. Nearly all electric street cars (Fig. 232), are set in motion by powerful motors placed under the cars. As the armature rotates, its motion is communicated by gears to the wheels, the necessary current reaching the motor through the overhead wires. Small motors may be used to great advantage in the home, where they serve to turn the wheels of sewing machines, and to operate washing machines. Vacuum cleaners are frequently run by motors.

CHAPTER XXXIV

HOW ELECTRICITY MAY BE MEASURED

CHAPTER XXXIV 241

312. **Danger of an Oversupply of Current.** If a small toy motor is connected with one cell, it rotates slowly; if connected with two cells, it rotates more rapidly, and in general, the greater the number of cells used, the stronger will be the action of the motor. But it is possible to send too strong a current through our wire, thereby interfering with all motion and destroying the motor. We have seen in Section 288 that the amount of current which can safely flow through a wire depends upon the thickness of the wire. A strong current sent through a fine wire has its electrical energy transformed largely into heat; and if the current is very strong, the heat developed may be sufficient to burn off the insulation and melt the wire itself. This is true not only of motors, but of all electric machinery in which there are current-bearing wires. The current should not be greater than the wires can carry, otherwise too much heat will be developed and damage will be done to instruments and surroundings.

The current sent through our electric stoves and irons should be strong enough to heat the coils, but not strong enough to melt them. If the current sent through our electric light wires is too great for the capacity of the wires, the heat developed will injure the wires and may cause disastrous results. The overloading of wires is responsible for many disastrous fires.

The danger of overloading may be eliminated by inserting in the circuit a fuse or other safety device. A fuse is made by combining a number of metals in such a way that the resulting substance has a low melting point and a high electrical resistance. A fuse is inserted in the circuit, and the instant the current increases beyond its normal amount the fuse melts, breaks the circuit, and thus protects the remaining part of the circuit from the danger of an overload. In this way, a circuit designed to carry a certain current is protected from the danger of an accidental overload. The noise made by the burning out of a fuse in a trolley car frequently alarms passengers, but it is really a sign that the system is in good working order and that there is no danger of accident from too strong a current.

313. **How Current is Measured.** The preceding Section has shown clearly the danger of too strong a current, and the necessity for limiting the current to that which the wire can safely carry. There are times

when it is desirable to know accurately the strength of a current, not only in order to guard against an overload, but also in order to determine in advance the mechanical and chemical effects which will be produced by the current. For example, the strength of the current determines the thickness of the coating of silver which forms in a given time on a spoon placed in an electrolytic bath; if the current is weak, a thin plating is made on the spoon; if the current is strong, a thick plating is made. If, therefore, the exact value of the current is known, the exact amount of silver which will be deposited on the spoon in a given time can be definitely calculated.

[Illustration: FIG. 233.--The principle of the galvanometer.]

Current-measuring instruments, or galvanometers, depend for their action on the magnetic properties of current electricity. The principle of practically all galvanometers is as follows:--

A closely wound coil of fine wire free to rotate is suspended as in Figure 233 between the poles of a strong magnet. When a current is sent through the coil, the coil becomes a magnet and turns so that its faces will be towards the poles of the permanent magnet. But as the coil turns, the suspending wire becomes twisted and hinders the turning. For this reason, the coil can turn only until the motion caused by the current is balanced by the twist of the suspending wire. But the stronger the current through the coil, the stronger will be the force tending to rotate the coil, and hence the less effective will be the hindrance of the twisting string. As a consequence, the coil swings farther than before; that is, the greater the current, the farther the swing. Usually a delicate pointer is attached to the movable coil and rotates freely with it, so that the swing of the pointer indicates the relative values of the current. If the source of the current is a gravity cell, the swing is only two thirds as great as when a dry cell is used, indicating that the dry cell furnishes about 1-1/2 times as much current as a gravity cell.

314. Ammeters. A galvanometer does not measure the current, but merely indicates the relative strength of different currents. But it is desirable at times to measure a current in units. Instruments for measuring the strength of currents in units are called ammeters, and

CHAPTER XXXIV

the common form makes use of a galvanometer.

A current is sent through a movable coil (the field magnet and coil are inclosed in the case) (Fig. 234), and the magnetic field thus developed causes the coil to turn, and the pointer attached to it to move over a scale graduated so that it reads current strengths. This scale is carefully graduated by the following method.

If two silver rods (Fig. 208) are weighed and placed in a solution of silver nitrate, and current from a single cell is passed through the liquid for a definite time, we find, on weighing the two rods, that one has gained in weight and the other has lost. If the current is allowed to flow twice as long, the amount of silver lost and gained by the electrodes is doubled; and if twice the current is used, the result is again doubled.

As a result of numerous experiments, it was found that a definite current of electricity will deposit a definite amount of silver in a definite time, and that the amount of silver deposited on an electrode in one second might be used to measure the current of electricity which has flowed through the circuit in one second.

A current is said to be one ampere strong if it will deposit silver on an electrode at the rate of 0.001118 gram per second.

[Illustration: FIG. 234.--An ammeter.]

In marking the scale, an ammeter is placed in the circuit of an electrolytic cell and the position of the pointer is marked on the blank card which lies beneath and which is to serve as a scale (Fig. 235). After the current has flowed for about an hour, the amount of silver which has been deposited is measured. Knowing the time during which the current has run, and the amount of deposit, the strength of the current in amperes can be calculated. This number is written opposite the place at which the pointer stood during the experiment.

The scale may be completed by marking the positions of the pointer when other currents of known strength flow through the ammeter.

[Illustration: FIG. 235.--Marking the scale of an ammeter.]

All electric plants, whether for heating, lighting, or for machinery, are provided with ammeters, such instruments being as important to an electric plant as the steam gauge is to the boiler.

315. **Voltage and Voltmeters.** Since electromotive force, or voltage, is the cause of current, it should be possible to compare different electromotive forces by comparing the currents which they produce in a given circuit. But two voltages of equal value do not give equal currents unless the resistances met by the currents are equal. For example, the simple voltaic cell and the gravity cell have approximately equal voltages, but the current produced by the voltaic cell is stronger than that produced by the gravity cell. This is because the current meets more resistance within the gravity cell than within the voltaic cell. Every cell, no matter what its nature, offers resistance to the flow of electricity through it and is said to have internal resistance. If we are determining the voltages of various cells by a comparison of the respective currents produced, the result will be true only on condition that the resistances in the various circuits are equal. If a very large external resistance of fine wire is placed in circuit with a gravity cell, the *total* resistance of the circuit (made up of the relatively small resistance in the cell and the larger resistance in the rest of the circuit) will differ but little from that of another circuit in which the gravity cell is replaced by a voltaic cell, or any other type of cell.

With a high resistance in the outside circuit, the deflections of the ammeter will be small, but such as they are, they will fairly accurately represent the electromotive forces which produce them.

Voltmeters (Fig. 236), or instruments for measuring voltage, are like ammeters except that a wire of very high resistance is in circuit with the movable coil. In external appearance they are not distinguishable from ammeters.

[Illustration: FIG. 236.--A voltmeter.]

The unit of electromotive force is called the *volt*. The voltage of a dry cell is approximately 1.5 volts, and the voltage of a voltaic cell and of a gravity cell is approximately 1 volt.

CHAPTER XXXIV

316. Current, Voltage, Resistance. We learned in Section 287 that the strength of a current increases when the electromotive force increases, and diminishes when the electromotive force diminishes. Later, in Section 288, we learned that the strength of the current decreases as the resistance in circuit increases.

The strength of a steady current depends upon these two factors only, the electromotive force which causes it and the resistance which it has to overcome.

317. Resistance. Since resistance plays so important a rôle in electricity, it becomes necessary to have a unit of resistance. The practical unit of resistance is called an ohm, and some idea of the value of an ohm can be obtained if we remember that a 300-foot length of common iron telegraph wire has a resistance of 1 ohm. An approximate ohm for rough work in the laboratory may be made by winding 9 feet 5 inches of number 30 copper wire on a spool or arranging it in any other convenient form.

In Section 299 we learned that substances differ very greatly in the resistance which they offer to electricity, and so it will not surprise us to learn that while it takes 300 feet of iron telegraph wire to give 1 ohm of resistance, it takes but 39 feet of number 24 copper wire, and but 2.2 feet of number 24 German silver wire, to give the same resistance.

NOTE. The number of a wire indicates its diameter; number 30, for example, being always of a definite fixed diameter, no matter what the material of the wire.

If we wish to avoid loss of current by heating, we use a wire of low resistance; while if we wish to transform electricity into heat, as in the electric stove, we choose wire of high resistance, as German silver wire.

CHAPTER XXXV

HOW ELECTRICITY IS OBTAINED ON A LARGE SCALE

318. **The Dynamo.** We have learned that cells furnish current as a result of chemical action, and that the substance usually consumed within the cell is zinc. Just as coal within the furnace furnishes heat, so zinc within the cell furnishes electricity. But zinc is a much more expensive fuel than coal or oil or gas, and to run a large motor by electricity produced in this way would be very much more expensive than to run the motor by water or steam. For weak and infrequent currents such as are used in the electric bell, only small quantities of zinc are needed, and the expense is small. But for the production of such powerful currents as are needed to drive trolley cars, elevators, and huge machinery, enormous quantities of zinc would be necessary and the cost would be prohibitive. It is safe to say that electricity would never have been used on a large scale if some less expensive and more convenient source than zinc had not been found.

319. **A New Source of Electricity.** It came to most of us as a surprise that an electric current has magnetic properties and transforms a coil into a veritable magnet. Perhaps it will not surprise us now to learn that a magnet in motion has electric properties and is, in fact, able to produce a current within a wire. This can be proved as follows:--

[Illustration: FIG. 237.--The motion of a magnet within a coil of wire produces a current of electricity.]

Attach a closely wound coil to a sensitive galvanometer (Fig. 237); naturally there is no deflection of the galvanometer needle, because there is no current in the wire. Now thrust a magnet into the coil. Immediately there is a deflection of the needle, which indicates that a current is flowing through the circuit. If the magnet is allowed to remain at rest within the coil, the needle returns to its zero position, showing that the current has ceased. Now let the magnet be withdrawn from the coil; the needle is deflected as before, but the deflection is in the opposite direction, showing that a current exists, but that it flows in the opposite direction. We learn, therefore, that a current may be induced in a coil by moving a magnet back and forth within the coil, but that a

CHAPTER XXXV

magnet at rest within the coil has no such influence.

An electric current transforms a coil into a magnet. A magnet in motion induces electricity within a coil; that is, causes a current to flow through the coil.

A magnet possesses lines of force, and as the magnet moves toward the coil it carries lines of force with it, and the coil is cut, so to speak, by these lines of force. As the magnet recedes from the coil, it carries lines of force away with it, this time reducing the number of the lines which cut the coil.

[Illustration: FIG. 238.--As long as the coil rotates between the poles of the magnet, current flows.]

320. **A Test of the Preceding Statement.** We will test the statement that a magnet has electric properties by another experiment. Between the poles of a strong magnet suspend a movable coil which is connected with a sensitive galvanometer (Fig. 237). Starting with the coil in the position of Figure 228, when many lines of force pass through it, let the coil be rotated quickly until it reaches the position indicated in Figure 238, when no lines of force pass through it. During the motion of the coil, a strong deflection of the galvanometer is observed; but the deflection ceases as soon as the coil ceases to rotate. If, now, starting with the position of Figure 238, the coil is rotated forward to its starting point, a deflection occurs in the opposite direction, showing that a current is present, but that it flows in the opposite direction. So long as the coil is in motion, it is cut by a varying number of lines of force, and current is induced in the coil.

The above arrangement is a dynamo in miniature. By rotation of a coil (armature) within a magnetic field, that is, between the poles of a magnet, current is obtained.

In the *motor*, current produces motion. In the *dynamo*, motion produces current.

321. **The Dynamo.** As has been said, the arrangement of the preceding Section is a dynamo in miniature. Every dynamo, no matter

how complex its structure and appearance, consists of a coil of wire which can rotate continuously between the poles of a strong magnet. The mechanical devices to insure easy rotation are similar in all respects to those previously described for the motor.

[Illustration: FIG. 239.--A modern electrical machine.]

The current obtained from such a dynamo alternates in direction, flowing first in one direction and then in the opposite direction. Such alternating currents are unsatisfactory for many purposes, and to be of service are in many cases transformed into direct currents; that is, current which flows steadily in one direction. This is accomplished by the use of a commutator. In the construction of the motor, continuous *motion* in one direction is obtained by the use of a commutator (Section 310); in the construction of a dynamo, continuous *current* in one direction is obtained by the use of a similar device.

322. Powerful Dynamos. The power and efficiency of a dynamo are increased by employing the devices previously mentioned in connection with the motor. Electromagnets are used in place of simple magnets, and the armature, instead of being a simple coil, may be made up of many coils wound on soft iron. The speed with which the armature is rotated influences the strength of the induced current, and hence the armature is run at high speed.

[Illustration: FIG. 240.--Thomas Edison, one of the foremost electrical inventors of the present day.]

A small dynamo, such as is used for lighting fifty incandescent lamps, has a horse power of about 33.5, and large dynamos are frequently as powerful as 7500 horse power.

323. The Telephone. When a magnet is at rest within a closed coil of wire, as in Section 319, current does not flow through the wire. But if a piece of iron is brought near the magnet, current is induced and flows through the wire; if the iron is withdrawn, current is again induced in the wire but flows in the opposite direction. As iron approaches and recedes from the magnet, current is induced in the wire surrounding the magnet. This is in brief the principle of the telephone. When one

CHAPTER XXXV

talks into a receiver, *L*, the voice throws into vibration a sensitive iron plate standing before an electromagnet. The back and forth motion of the iron plate induces current in the electromagnet *c*. The current thus induced makes itself evident at the opposite end of the line *M*, where by its magnetic attraction, it throws a second iron plate into vibrations. The vibrations of the second plate are similar to those produced in the first plate by the voice. The vibrations of the far plate thus reproduce the sounds uttered at the opposite end.

[Illustration: FIG. 241.--Diagram of a simple telephone circuit.]

324. Cost of Electric Power. The water power of a stream depends upon the quantity of water and the force with which it flows. The electric power of a current depends upon the quantity of electricity and the force under which it flows. The unit of electric power is called the watt; it is the power furnished by a current of one ampere with a voltage of one volt.

One watt represents a very small amount of electric power, and for practical purposes a unit 1000 times as large is used, namely, the kilowatt. By experiment it has been found that one kilowatt is equivalent to about 1-1/3 horse power. Electric current is charged for by the watt hour. A current of one ampere, having a voltage of one volt, will furnish in the course of one hour one watt hour of energy. Energy for electric lighting is sold at the rate of about ten cents per kilowatt hour. For other purposes it is less expensive. The meters commonly used measure the amperes, volts, and time automatically, and register the electric power supplied in watt hours.

INDEX

Absorption, of heat by lampblack, 143-144. of gases by charcoal, 57. of light waves, 135-138.

Accommodation of the eye, 123.

Acetanilid, 259.

Acetylene, as illuminant, 152-153. manufacture of, 152-153. properties of, 220.

Acid, boric, 253. carbolic, 152, 251, 252. hydrochloric, 55, 80, 227, 238, 241. lactic, 230. oxalic, 247, 248. salicylic, 253. sulphuric, 55, 80, 240, 241, 307. sulphurous, 242.

Acids, action on litmus, 220.

Adenoids, 51.

Adulterants, detection of, 16.

Air, characteristics of, 81-83, 86, 189. compressibility of, 91. expansion of, 10-11. humidity, 38, 39. pumps, 201-205. transmits sound, 269. weight of, 86. *See* Atmosphere.

Alcohol, 234. in patent medicines, 260.

Alizarin, 248.

Alkali, 222.

Alternating current, 351.

Alum, 247. in baking powder, 230.

Ammeter, 341, 343.

Ammonia, 152. a base, 221-222. in bath, 226. in manufacture of ice, 98. neutralizing chlorine, 240.

Ampere, 342.

Anemia, 259.

Angle, of incidence, 110. of reflection, 110. of refraction, 114.

Aniline, 152, 245.

Animal charcoal, 58.

Animal transportation, 132.

Antichlor, 240.

Antipyrin, 259.

Armature, 319, 320. dynamo, 350. motor, 335.

Artificial lighting, 148-153.

Atmosphere, 81. carbon dioxide in, 54-55. height of, 81. nitrogen and oxygen in, 262. pressure of, 82-86. water vapor in, 36-38. weight, 86. *See* Air.

Atmospheric pressure, 82-86.

Atomizer, 92.

Atoms, 102.

Automobiles, gas engines, 185.

Axis of a lens, 119.

Bacteria, 133. as nitrogen makers, 263. destroyed by sunlight, etc., 133, 250, 251. diseases caused by, 133. in butter and cheese, 133.

Baking powder, 229-230.

Baking soda, 227-229.

Barograph, 87.

Barometer, aneroid, 84-85. mercury, 84. use in weather predictions, 86-87.

Bases, action on litmus, 221-222. properties, 220-222.

Battery, electric, 311.

Beans, as food, 66. roots take in nitrogen, 263.

Bell, electric, 319-321.

Benzine, 150. as a cleaning agent, 227.

Benzoate of soda, 253.

Bicarbonate of soda, in fire extinguisher, 55, 56. in Rochelle salt, 227. in soda mints, 231. in seidlitz powder, 231.

Bicycle pumps, 202.

Blasting, by electricity, 314.

Bleaching, 237-243. by chlorine, 238-240.

Bleaching powder, 239-240.

Body, human, 63-64. a conductor of electricity, 292.

Boiling, 31. amount of heat absorbed, 31-32. of milk, 32. of water, 77. point, 15.

Bomb calorimeter, 61.

Borax, as meat preservative, 253. as washing powder, 226.

Boric acid, as meat preservative, 253.

Boyle's law, 95-96.

Bread, 232-233. unleavened, 233.

Bread making, 232-235.

Breathing, hygienic habits of, 50. by mouth, 50-51.

Burns, treatment of, 52-53.

Butter, adulteration test, 16. bacteria in, 133.

Buttermilk, 230.

Caisson, 203-204.

Calcium carbide, 152-153. in making nitrogenous fertilizer, 264.

Calico printing, 249.

Calorie, 27-28, 61-62.

Calorimeter, 61.

Camera, 128-129. films, 129. lens, 129. plates, 129.

Camping, water supply, 195-197.

Candle, 148-149. as standard for light-measure, 104-105.

Candle-power, 105-107.

Carbide, calcium, 152-153, 264.

Carbohydrates, 64-65, 149.

Carbolic acid, 152. as disinfectant, 251.

Carbon, 56, 66. in voltaic cells, 308.

Carbon dioxide, 53. as fire extinguisher, 55-56. commercial use, 55-56. in baking soda, 228. in fermentation, 234. in health, 54. in plants, 55. preparation of, 55. source of, 53. test for, 228.

Catarrh, 259.

Caustic lime, 222..

Caustic potash, 222.

Caustic soda, 218, 222. to make a salt, 227.

Caves and caverns, 71.

Cell, dry, 310. gravity, 309-310. voltaic, 306-308, 310.

Cells of human body, 63, 64, 66.

Centigrade thermometer, 15.

Central heating plant, 19.

Chalk, in making carbon dioxide, 55.

Charcoal as a filter, 57. commercially, 57. preparation, 57-58.

Chemical action, and electricity, 307, 315-317. and light, 126, 127.

Chemistry, in daily life, 218, 219.

Chills, 38.

Chloride of lime, in bleaching, 240. disinfectant, 251.

Chlorine, and hydrogen, 239. effect upon human body, 239. in bleaching, 238-240. influence of light upon, 126. presence in salt, 227.

Circuit, electric, 321. local, in telegraph, 325-326.

City water supply, 206-212.

Clarinet, 297.

Cleaning of material, 226, 243.

Climate, influenced by presence of water, 29, 40.

Clover, nitrogen producers, 263.

Coal, 30.

Coal gas, 150, 151. by-products, 152.

Coal oil, 149, 150.

Coal tar dyes, 152, 218, 245.

Cogwheels, 170.

Coil, current-bearing, 320. magnetic field about, 331-333.

Coke, 152.

Cold storage, 97.

Color, 134-141. and heat, 142, 143. influenced by light, 137. of opaque bodies, 136, 137. of transparent bodies, 135, 136.

Color blindness, 140, 141. designs in cloth, 248, 249.

Colors, compound, 138, 139. essential, 139-140. primary, 135. simple, 138. spectrum, 134-135. variety in dyeing, 247, 248.

Combustion, heat of, 45. spontaneous, 52.

Commutator, 335.

Compass, 328.

Compound colors, 138, 139.

Compound machine, 171.

Compound substances, 103.

Compression of air, 91, 92. cause of heat, 96.

Compression pumps, 201, 205.

Concave lens, 118.

Condensation, 33. heat set free, 40.

Conduction of heat, 25.

Conductivity metals, 321.

Conductors, electric, 321, 322.

Conservation, of energy, 58, 59. of matter, 58, 59.

Convection, 24, 25.

Convex lens, 118.

Cooling, by evaporation, 35-36. by expansion, 97.

Copper, in electric cell, 307.

Core, iron, 319.

Corn, bleached with sulphurous acid, 242.

Cotton, mercerized, 218. bleaching, 241. dyeing, 245-247.

Cough sirup, 258.

Crane, compound machine, 172.

Cream of tartar, 229.

Creosote oil, 254.

Crude petroleum, 149, 150.

Current, electric, 306, 312. alternating, 349. induced, 346-347. measurement of, 340. resistance, 312, 343, 345. strength, 339, 340, 344.

Dams, 214-216.

Decay, 49.

Decomposition of soil by water, 70-74.

Degrees Fahrenheit and Centigrade, 15.

Density, 11.

Designs in cloth, printed, 248, 249. woven, 249.

Developer in photography, 128.

Dew, 36, 37.

Dew point, 38.

Diarrhea, 251.

Diet, 62, 66. economy on table, 66-69.

Discord, reason for, 271.

Disease, and surface water, 76. relation of light to, 131-132.

Disease disinfectants, 250, 251, 252.

Distillation, 34-35. in commerce, 35. of petroleum, 149-150. of soft coal, 150. of water, 34, 35, 77.

Diving suits, 204.

Door bells, 319-321.

CHAPTER XXXV

Drainage, of land, 194, 195. sewage, 196, 198, 199, 201.

Drilled well, 199.

Drinking water, 75-77. in camping, 195-196. and rural supplies, 198, 201.

Driven well, 196-197.

Drought, 217.

Drugs, 255, 260.

Dry cell, 312.

Dyeing, 244-249. color designs, 248.

Dyeing, direct, 245. home, 247. indirect, 247. variety of color, 247.

Dyes, 218, 244, 245.

Dynamo, 346. alternating current, 349. source of energy, 346-347.

Ear, in man, 301-303. care of, 303.

Earth, conductor of electricity, 326.

Echo, 277.

Economy in buying food, 66-69.

Effort, muscular, 155, 160.

Electric, battery, 311. bell, 319-321. bread toasters, 314. conductors and non-conductors, 321-322. cost of, energy, 352. current, 306, 312. flatiron, 313. heating pad, 314. lights, 314. street cars, 337.

Electricity, heat, 312-315, 339. as a magnet, 319, 331-333. practical uses of, 312-317.

Electrodes, of cell, 308.

Electrolytic metals, 317.

Electromagnets, 319.

Electromotive force, 308. unit of, 344.

Electroplating, 315.

Electrotyping, 317.

Elements, 102-103.

Emulsion, 224.

Energy, conservation of, 58, 59. transformations of, 58, 59.

Engine, steam, 183-185. gas, 185-186. horse power, 173.

Erosion, 73-74.

Essential colors, 139-140.

Evaporation, 35-39. cooling effect, 35-36. effect of temperature on, 35, 36. effect of air on, 38. freezing by, 98. heat absorbed, 36. of perspiration, 38.

Expansion, of air, 10, 11. cooling effect of, 97. disadvantage and advantage of, 11-13. of liquids, 9-11. of solids, 10, 11. of water, 9, 10, 11, 12. Eye, 122-125. headache, 124, 125. how focused, 122, 123. nearsighted and farsighted, 123. strain, 125.

Fahrenheit thermometer, 15.

Fats, 65. in soap making, 223.

Fermentation, 232-236. by yeast, 234-236.

Ferric compounds, 248.

Fertilizers, 262-265. nitrogen, 262. phosphorus, 263, 264. potash, 263-265.

Field magnet, 336.

Filings, iron, 329.

Film, photographic, 129.

Filter, charcoal, 57.

Filtering water, 77.

Fire, 9. and oxygen, 45, 47. and tinder box, 47. making of, 51. primitive production of, 47. produced by friction, 47. spontaneous combustion, 52. sores and burns, 52-53. extinguisher, 55, 56.

Fireless cooker, 25, 26.

Fireplaces, 17, 18.

Fixing, in photography, 128.

Flame, hydrogen, 80.

Flood, Johnstown, 214, 215. relation to forests, 217.

Flour, self-raising, 231.

Flume, 177.

Flute, 297.

Focal length, 118.

Focus, of lens, 118.

CHAPTER XXXV

Fog, 37.

Food, 60-69. carbohydrates, 64, 65. economy in buying, 66-69. fats, 65. fuel value of, 60-62. need of, 63, 64. preservatives, 252. proteids, 66. value, 67. waste, 60. water in, 75.

Foot pound, 172.

Force and motion, 156, 157. and work, 156, 157. magnetic lines of, 329-331, 334. muscular, 155, 160.

Force pumps, 192, 193.

Forests and water supply, 216-217.

Forging of iron, 40, 41.

Formaldehyde, 253.

Freezing, effect of salt, 44. effect on ground and rocks, 42. expansion of water on, 41. ice cream freezer, 44.

Frequency in music, 273, 275.

Fresh air, 22-24, 49. amount consumed by gas burner, 22. and health, 49, 50. in underground work, 202. in work under water, 203-205.

Friction, 173, 174. losses by, 174, 210. source of heat and fire, 47.

Frost, 36, 37.

Fruit, canned, bleached with sulphurous acid, 242. colored with coal tar dyes, 253.

Fuel value of foods, 60-62. table of fuel values, 67.

Fulcrum, 159, 160.

Fumigation, 251.

Fundamental tone, 290, 291, 292.

Furnace, hot air, 19.

Fuse, 340.

Fusion, heat of, 40.

Galvanometer, 341.

Gas, acetylene, 152, 153. and unburned carbon, 151. coal, 151, 152. effect of heat on volume, 96, 97. effect of pressure on volume, 95-96. engine, 185-186. for cooking, 151, 152. illuminating, 92, 93, 150, 151. liquefaction, 97, 98. meter, 93, 94. natural, 152.

Gasolene, 149, 150. as cleaning agent, 227, 243. in gas engine, 185, 186.

Gauge, pressure, 92-94.

Gelatin, plate and film, 129.

Glass, kinds of, 119. molding of, 40. non-conductor, 321.

Grape juice, fermented with millet, 233.

Gravity cell, 309, 310.

Grease, and lye, 221. and soap making, 223.

Gulf Stream, 24.

Hard water, and soap, 225.

Harp, 295.

Headache, 124, 125. powders, 259.

Health, effect of diet, 62, 64.

CHAPTER XXXV

Heat, 9. absorbed in boiling, 31-32. and disease germs, 250. and food, 252. and friction, 47. and light, 142, 147. and oxidation, 45, 48, 49. and wave motion, 145-147. conduction, 25. convection, 24, 25. from burning hydrogen, 80. from electricity, 312-315, 339. needed to melt substances, 39. of fusion, 40. of vaporization, 32. produced by compression, 96. relation of water to weather, 29, 40. set free by freezing water, 40. sources of, 29-30. specific, 28-29. temperature, 27. unit of, 27, 28.

Heating effect of electric current, 312-315.

Heating of buildings: central heating plant, 19. fireplaces, 17-18.

Heating, furnaces, 19. hot water, 19-22.

Helix, 318.

Horse power, 173, 351.

Hot water heating, 19-22.

Hues, primary, 135.

Humidity, 38. proper percentage for health and comfort, 38, 39.

Humus, 216, 217.

Hydrocarbons, 149.

Hydrochloric acid, composition, 227. in bleaching, 241. to make a salt, 227. to make carbon dioxide, 55. to make chlorine, 238. to make hydrogen, 80.

Hydrogen, 65, 66. and chlorine, 239. and water, 79. chemical conduct, 126-127. flame, 80. in voltaic cell, 307. peroxide, 53, 252. preparation, 80. to liquefy, 97.

Ice, lighter than water, 42. manufacture of, 98, 99.

CHAPTER XXXV

Ice cream freezers, 44.

Illuminating gas, manufacture of, 150, 151. measurement of quantity consumed, 93, 94. test of pressure, 92, 93.

Illumination, intensity of, 105, 106.

Image, in mirror, 108, 111.

Incandescent lighting, 107, 314.

Incidence, angle of, 110.

Inclined plane, 162-166. screw, 166. wedge, 166.

Indigo, 218.

Induced current, 346-347.

Ink spots, removal of, 243.

Insoluble substances, 71.

Insulators, electric, 324.

Intensity, of light, 105-107. of sound, 270-271.

Interval, in musical scale, 283.

Iron, forging, 41. filings, 329. galvanizing, 49. oxidation of, 48.

Irrigation, 193-194.

Isobaric lines, 88, 91.

Isothermal lines, 89, 91.

Johnstown flood, 214, 215.

CHAPTER XXXV

Kerosene, 149, 150.

Kilowatt, 351.

Lactic acid, 230.

Leaves, 132, 262.

Lens, 117-121. concave, 118. converging, 118. crystalline, of eye, 122. focal length, 118. material, 119. refractive power, 119.

Lever, 158-162. examples, 160-162. fulcrum, 159, 160.

Life, and carbon dioxide, 54. and nitrogen, 261. and oxygen, 49, 54.

Lifting pumps, 189-192.

Light, absorption, 135-138. and heat, 142-147. a wave motion, 145-147. bent rays, 113, 114. chemical action, 126-127. disease, 131-132. essential to life, 131, 132. fading illumination, 105, 106. influence on color, 134. reflection of, 109-112. refraction of, 113-125. travels in a straight line, 108. white, composed of colors, 134.

Lighting, artificial, 148-153.

Lime, chloride of, 240, 251.

Limewater, 220. and carbon dioxide, 228.

Linen, bleaching, 241. dyeing, 245-247.

Lines, of force, 329-331, 334. isobaric, 88, 91. isothermal, 89, 91.

Liquefaction of gases, 97, 98.

Liquid air, 98.

Liquid soap, 223, 224.

Litmus, action of acids, 220. action of bases, 221, 222. action of neutral substance, 222.

Logwood dyes, 245, 247, 248.

Los Angeles aqueduct, 211.

Lye, 221, 222.

Machines, compound, 171. inclined plane, 162-166. lever, 158-162. pulley, 166-169. wheel and axle, 169-171.

Madder, for dyes, 245.

Magnet, 328. electro-, 319. field of, 329-331. lines of force about, 329-331. poles of, 330-332. properties of electricity, 318.

Magnetic, needle, 328. poles, 329-331.

Magnifying power, of a lens, 115. of a microscope, 115. of a telescope, 115.

Mammoth Cave of Kentucky, 71.

Manganese dioxide, 46. chlorine made from, 238. oxygen made from, 46.

Marble, for carbon dioxide, 55.

Matches, 47. safety, 47-48.

Matching colors, 137.

Matter, conservation of, 58, 59.

Meat, 66. preservation of, 253.

Mechanical devices, 154, 155.

Melting, 39, 40.

Melting point, 40.

Melting substances without a definite melting point, 40.

Mercerized cotton, 218.

Mercury, barometer, 84. thermometer, 14-17.

Metals, electroplating, 317. preservation by paint, 253-254. veins deposited by precipitation, 72, 73. welding, 315.

Meter, gas, 93, 94.

Microörganisms, 132, 133.

Microscope, 115.

Milk, boiling point, 32. Pasteurized, 250.

Minerals, in foods, 62, 63. in water, 70, 71.

Mirrors, 108-112. distance of image behind mirror, 111. distance of object in front of mirror, 111. image a duplicate of object. 111.

Molding of glass, 40.

Molecule, 100-103.

Mordants, 247, 248, 249.

Morphine, 257.

Morse, telegraphic code, 324.

Motion, in sound, 266, 278, 280. in work, 156.

Motor, electric, 336. principle of, 333. street car, 337.

Mouth breathing, 50. cause of, 51.

Movable pulley, 167, 168.

Music, 278.

Musical instruments, percussion, 299. stringed, 284-295. wind, 295, 299.

Musical scale, 282.

Naphtha in gas engines, 185.

Naphthalene, 152.

Narcotics, 255.

Natural gas, 152.

Needle, magnetic, 328.

Negative, electrode, 308. photographic, 130.

Neutral substance, 222. and litmus, 222.

Neutralization, 222.

Niagara Falls, 176.

Nitrogen, 66. and bacteria, 263. and plant life, 261. in atmosphere, 261. in fertilizer, 262-265. in food, 66. preparation of, 261. properties of, 261.

Noise in music, 280.

Non-conductors, of electricity, 321-322. of heat, 25.

Nutcracker, as a lever, 162.

CHAPTER XXXV 269

Oboe, 297.

Octave, 284.

Odors, 101.

Ohm, unit of resistance, 345.

Oil, gasoline, 149, 150. kerosene, 149, 150. lubricating, 174. olive, 16.

Orchestra grouping, 299.

Ore, 72.

Organ pipes, 297.

Overtones, 290-293.

Oxalic acid, 247, 248.

Oxidation, 45-59. and decay, 49. heat the result of, 49-52. in human body, 49, 53. of iron, 48.

Oxygen, 66. and bleaching, 239. and combustion, 45. and food, 66. and plants, 55. and the human body, 50. and water, 79, 80. in the atmosphere, 45. preparation of, 46.

Paint, as wood and metal preservatives, 253, 254. removal of stains, 243.

Paper making, 219.

Paraffin, 150, 321.

Pasteurized milk, 250.

Patent medicines, 257-260.

Peas, sources of nitrogen, 263.

Pelton wheel, 177.

Percussion instruments, 299.

Period of a body, 273.

Peroxide of hydrogen, 53, 252

Petrolatum, 150.

Petroleum, 149, 150.

Phonograph, 303-305.

Phosphorus, in fertilizer, 263, 264. in making nitrogen, 261. in matches, 47, 48. poisoning by, 47.

Photography, 127-131.

Photometer, 107.

Pianos, 284-292.

Pin wheel, 181.

Pitch of sound, 280, 281. cause of, 282. in wind instruments, 296-299.

Plane, inclined, 162-166.

Plants, and atmosphere, 55. and light, 131-132. and nitrogen, 261.

Plate developing, photographic, 128.

Pneumatic dispatch tube, 205.

Poles, magnetic, 330-332. of cell, 308.

Positive electrode, 308.

CHAPTER XXXV

Potash, in fertilizer, 263-265.

Potassium chlorate and oxygen, 46. permanganate, 100. tartrate and Rochelle salt, 227.

Power, candle, 105-107. electric, 351. horse, 173, 351. sources of, 174, 175, 185. transmission by belts, 171. water, 176-180.

Precipitation, 72, 73.

Preservatives, food, 252. wood and metal, 253-254.

Pressure, atmospheric, 82-86. calculation of atmospheric, 83, 84. calculation of gas, 92, 93. calculation of water, 94. gauge, 92-94. of illuminating gas, 93. relation of pressure of gas to volume, 95, 96. water pressure, 208-211, 214-216. within the body, 86.

Primary colors, 135.

Print, photographic, 131.

Printing, color designs in cloth, 248, 249. electrotype, 317.

Prisms, 135. refraction through, 117.

Proteids, 66.

Pulleys, 166-169. applications of, 169.

Pump, 187-205. air, 201-205. force, 192, 193. lifting, 189-192.

Pupil of the eye, 122.

Pure food laws, bleaching, 242. preservatives, 252.

Purification of water, 77, 196.

Push button, 321.

CHAPTER XXXV

Radiator, 19-21.

Railroads, grading of, 165-166.

Rain, 36, 37.

Rainbow, 134.

Rain water, 225.

Reflection, angle of, 110. of light, 109-112. of sound, 278, 279.

Refraction, angle of, 114. by atmosphere, 114. of light, 113. uses of, 115-116.

Relay, telegraph, 325.

Reservoir, 214. artificial, 211. construction of, 214-216. natural, 211.

Resistance, electrical, 312. internal, of cell, 343. unit of, 345.

Resonance, 276.

River, volume and value of, 180.

Roads, application of inclined plane to, 165-166.

Rochelle salt, 227, 231.

Rocks, effect of freezing water on, 42-43. water as a solvent, 71.

Rosin, obtained by distillation, 35.

Safety matches, 47-48.

Salicylic acid, 253.

Salt, 227-228.

CHAPTER XXXV

Salts, 227. general properties, 227. in ocean, 227. smelling, 222.

Saturation of air, 37.

Scale, musical, 282.

Screw, and inclined plane, 166.

Seaweed, 265.

Seidlitz powder, 231.

Self-raising flour, 231.

Sewage, disposition of, 198-199. of camps, 196. source of revenue, 201.

Sewer gas, 57.

Silk, bleaching, 241. dyeing, 245-247.

Silver chloride, 127-131.

Simple colors, 138.

Simple substances, 103.

Siren, 280.

Smelling salts, 222.

Snow, 36-37.

Soap, 222-224. and hard water, 225. liquid, 223-224. preparation, 223.

Soda, baking, 227, 228-229. benzoate, 253. caustic, 218, 222, 223, 227. washing, 225, 226, 229.

Soda mints, 231.

Sodium, bicarbonate, 56, 227, 228, 230-231. carbonate, 228. chloride, 228.

Soil, deposited by streams, 73.

Solenoid, 318.

Solution, 70.

Soothing sirup, 258.

Sound, and motion, 266, 278. musical, 278. nature of, 266. reflection, 277. speed of, 271-272. transmission of, 267-271. velocity of, 271-272. waves, 272-274.

Sounder, telegraph, 324.

Sounding board, 277.

Sour milk in cooking, 230.

Specific heat, 28-29.

Spectrum, 134-135.

Speed, of sound, 271, 272.

Spontaneous combustion, 52.

Stains, removal of, 226, 243.

Standpipes, 212.

Starch, 65.

Steam, and work, 183-184. engine, 183-185. heat of vaporization, 32. heating by, 33. turbine, 183-184.

Steel, forging and annealing, 16.

CHAPTER XXXV

Stoves, 18-19.

Streams, carriers of mud, 73. volume of, 179-180.

Street cars, electric, 337.

Stringed instruments, 284-295.

Strings, vibrating, 286-290.

Sugar, 16, 65. fermented by yeast, 234.

Sulphur, 66. as disinfectant, 251. in making sulphurous acid, 242.

Sulphuric acid, in bleaching, 240,241. in fire extinguisher, 55. in making of hydrogen, 80. in voltaic cell, 307.

Sulphurous acid, in bleaching, 242. preparation, 242.

Sun, energy derived from, 143-144. source of heat, 29-30.

Sunlight, 135. and bacteria, 133. and chemical action, 126-127.

Sympathetic vibrations, 274-277.

Tallow, 105, 148.

Tartar, cream of, 229.

Telegraph, 322. long distance, 327. relay, 325. sounder, 324.

Telephone, 350-351.

Temperature, 13-14. as measurement of heat present, 27. in detecting adulterants, 17. in forging steel, 16. in making sirups, 16. measurement of, 14-15.

Thermometer, 14-17. Centigrade, 15. Fahrenheit, 15.

Tinder box, 47.

Transmission, of light, 145-147. of sound, 267-271.

Tuning fork, 266, 273, 278, 290.

Turbine, steam, 183. water, 178.

Turpentine, and grease, 226. by distillation, 35.

Unleavened bread, 233.

Vacuum, sound in, 268.

Vapor, in atmosphere, 36-38.

Vaporization, heat of, 32.

Varnish, on candies, 253.

Vegetable matter, and coal, 30. and gas, 30. and oil, 30.

Veins, formation in rock, 72-73.

Velocity, of sound, 271-272.

Ventilation, 21-24, 54. need of, 38.

Vibration, of strings, 286-290. sympathetic, 274-277.

Viola, 295.

Violin, 295.

Violoncello, 295.

Vocal cords, 300.

Voice, 300.

CHAPTER XXXV

Volt, 344.

Voltage, 345.

Voltaic cell, 306-308, 310.

Voltmeter, 344.

Volume, of a stream, 179-180. relation of pressure of a gas, 95-96.

Washing powders, 224-226. soda, 229.

Water, action in nature, 70-74. amount used daily per person, 181. and hydrogen, 79. and oxygen, 79, 80. as solvent, 70-71. boiling, 77. boiling point, 15. composition, 79-80. condensation, 33. dams and reservoirs, 214-216. density, 11. distilled, 34, 77. drinking, 75-77, 195-201. electrolysis, 79-80. evaporation, 33-34. expansion, 9-10, 41-42. filtration, 77. freezing, 40-41. hard, 225. heat of fusion, 40. impurities, 76-77. in atmosphere, 36-38. in food, 75. in human body, 75. in vegetables, 75. influence on climate, 29, 40. irrigation, 193-194. minerals in, 70-71. ocean, 265. power, 176-180. precipitates, 72, 73. pressure, 208-211, 214-216. purification, 77. rain, 225. running, value of, 178-180. source of, 78. steam, 32. waves, 145-147. weight, 208-209, 215. wells, 195-201. wheels, 176-180. work under, 203-205.

Water supply, and forests, 216-217. cost, 212-214. of city, 206-212, 217.

Watt, 351.

Waves, heat, 145-147. light, 145-147. sound, 268, 272-274. water, 145-147.

Weather, bureau, 87-91. forecasts, 38-39, 86-88. relation of water to, 29, 40.

Weather maps, 89-91.

Wedge, and inclined plane, 166.

Weight, of air, 86. of water, 208-209, 215.

Welding, by electricity, 315.

Wells, 195-201. drilled, 199. driven, 196-197.

Wheel and axle, 169-171. cogwheels, 170. windlass, 169.

Wheelbarrow as lever, 160-161.

White light, nature of, 135.

Wind instruments, 297-301.

Windlass, 169.

Windmill, 174-175, 180-182.

Winds, 24.

Wine, 232, 234.

Wood, as source of charcoal, 58. ashes in soap making, 223. in paper making, 219. preservation, 253-254.

Wool, bleaching, 241. dyeing, 245-247.

Work, 156-186. and steam, 183-184. and water, 176-180. conservation, 174-175. formula, 157. machines, 157-175. unit of, 172-173. waste, 173.

Woven designs in cloth, 249.

Yeast, 234-236. wild, 235-236.

Zinc, in galvanizing iron, 49. in making hydrogen, 80. in voltaic cell, 307-308.

PLANT LIFE AND PLANT USES

CHAPTER XXXV

By JOHN GAYLORD COULTER, Ph. D.

$1.20

An elementary textbook providing a foundation for the study of agriculture, domestic science, or college botany. But it is more than a textbook on botany--it is a book about the fundamentals of plant life and about the relations between plants and man. It presents as fully as is desirable for required courses in high schools those large facts about plants which form the present basis of the science of botany. Yet the treatment has in view preparation for life in general, and not preparation for any particular kind of calling.

The subject is dealt with from the viewpoint of the pupil rather than from that of the teacher or the scientist. The style is simple, clear, and conversational, yet the method is distinctly scientific, and the book has a cultural as well as a practical object.

The text has a unity of organization. So far as practicable the familiar always precedes the unfamiliar in the sequence of topics, and the facts are made to hang together in order that the pupil may see relationships. Such topics as forestry, plant breeding, weeds, plant enemies and diseases, plant culture, decorative plants, and economic bacteria are discussed where most pertinent to the general theme rather than in separate chapters which destroy the continuity. The questions and suggestions which follow the chapters are of two kinds; some are designed merely to serve as an aid in the study of the text, while others suggest outside study and inquiry. The classified tables of terms which precede the index are intended to serve the student in review, and to be a general guide to the relative values of the facts presented. More than 200 attractive illustrations, many of them original, are included in the book.

AMERICAN BOOK COMPANY

A NEW ASTRONOMY, $1.30

By DAVID TODD, M. A., Ph. D., Professor of Astronomy and Navigation and Director of the Observatory, Amherst College.

Astronomy is here presented as preeminently a science of observation. More of thinking than of memorizing is required in its study, and greater emphasis is laid on the physical than on the mathematical aspects of the science. As in physics and chemistry, the fundamental principles are connected with tangible, familiar objects, and the student is shown how he can readily make apparatus to illustrate them. In order to secure the fullest educational value, astronomy is regarded as an inter-related series of philosophic principles.

* * * * *

MATHEMATICAL GEOGRAPHY, $1.00

By WILLIS E. JOHNSON, Ph. D., Vice-President and Professor of Geography and Social Sciences, Northern Normal and Industrial School, Aberdeen, South Dakota.

This work explains with great clearness and thoroughness that portion of the subject which not only is most difficult to understand, but also underlies and gives meaning to all geographical knowledge. A vast number of facts which are much inquired about, but little known, are taken up and explained. Simple formulas are given so that a student unacquainted with geometry or trigonometry may calculate the heights and distances of objects, the latitude and longitude of a place, the amount any body is lightened by the centrifugal force due to rotation, the deviation of a plumb-line from a true vertical, etc.

AMERICAN BOOK COMPANY

ELEMENTS OF GEOLOGY

By ELIOT BLACKWELDER, Associate Professor of Geology, University of Wisconsin, and HARLAN H. BARROWS, Associate Professor of General Geology and Geography, University of Chicago.

$1.40

An introductory course in geology, complete enough for college classes, yet simple enough for high school pupils. The text is explanatory, seldom merely descriptive, and the student gains a knowledge not only of the salient facts in the history of the earth, but also of the methods by which those facts have been determined. The style is simple and direct. Few technical terms are used. The book is exceedingly teachable.

The volume is divided into two parts, physical geology and historical geology. It differs more or less from its predecessors in the emphasis on different topics and in the arrangement of its material. Factors of minor importance in the development of the earth, such as earthquakes, volcanoes, and geysers, are treated much more briefly than is customary. This has given space for the extended discussion of matters of greater significance. For the first time an adequate discussion of the leading modern conceptions concerning the origin and early development of the earth is presented in an elementary textbook.

The illustrations and maps, which are unusually numerous, really illustrate the text and are referred to definitely in the discussion. They are admirably adapted to serve as the basis for classroom discussion and quizzes, and as such constitute one of the most important features of the book. The questions at the end of the chapters are distinctive in that the answers are in general not to be found in the text. They may, however, be reasoned out by the student, provided he has read the text with understanding.

AMERICAN BOOK COMPANY

ESSENTIALS OF BIOLOGY

By GEORGE WILLIAM HUNTER, A. M., Head of Department of Biology, De Witt Clinton High School, New York City.

$1.25

This new first-year course treats the subject of biology as a whole, and meets the requirements of the leading colleges and associations of science teachers. Instead of discussing plants, animals, and man as separate forms of living organisms, it treats of fife in a comprehensive manner, and particularly in its relations to the progress of humanity. Each main topic is introduced by a problem, which the pupil is to solve by actual laboratory work. The text that follows explains and illustrates the meaning of each problem. The work throughout aims to have a human interest and a practical value, and to provide the simplest and most easily comprehended method of demonstration. At the end of each chapter are lists of references to both elementary and advanced books for collateral reading.

* * * * *

SHARPE'S LABORATORY MANUAL IN BIOLOGY

$0.75

In this Manual the 56 important problems of Hunter's Essentials of Biology are solved; that is, the principles of biology are developed from the laboratory standpoint. It is a teacher's detailed directions put into print. It states the problems, and then tells what materials and apparatus are necessary and how they are to be used, how to avoid mistakes, and how to get at the facts when they are found. Following each problem and its solution is a full list of references to other books.

AMERICAN BOOK COMPANY

ESSENTIALS OF PHYSICS

By GEORGE A. HOADLEY, C.E., Sc. D., Professor of Physics, Swarthmore College.

$1.25

This is the author's popular and successful Elements of Physics enriched and brought up to date. Despite the many changes and modifications made in this new edition, it retains the qualities which have secured so great a success for the previous book.

It tells only what everyone should know, and it does this in a straightforward, concise, and interesting manner. It takes into consideration the character of high school needs and conditions, and, throughout, lays particular emphasis upon the intimate relation between physics and everyday life.

While the subject matter, as a whole, is unchanged, the order of topics in many cases has been altered to adapt the development of the subject to the habits of thought of high school pupils. Instead of beginning the treatment of a subject with the definition and proceeding to a discussion of the sub-topics, the author starts with a discussion of well-known phenomena and leads up to the definition of the subject discussed. The text, wherever possible, has been simplified, more than fifty topics having been amplified, expanded, or reworded. More familiar illustrations of the topics treated are given, and the demonstrations of many of the experiments are simplified by the use of materials that are readily obtainable in the classroom.

There have been added a number of new topics, mostly in connection with the recent advances in applied science. The number both of questions and problems has been greatly increased and the data in these all relate to actual, practical, physical phenomena. More than one-fifth of the illustrations in the book are new, many of the pictures of apparatus having been redrawn to show modern forms.

AMERICAN BOOK COMPANY

End of the Project Gutenberg EBook of General Science, by Bertha M. Clark

*** END OF THIS PROJECT GUTENBERG EBOOK GENERAL SCIENCE ***

***** This file should be named 16593-8.txt or 16593-8.zip ***** This and all associated files of various formats will be found in:
http://www.gutenberg.org/1/6/5/9/16593/

Produced by John Hagerson, Kevin Handy, Sankar Viswanathan and the Online Distributed Proofreading Team at http://www.pgdp.net

Updated editions will replace the previous one--the old editions will be renamed.

Creating the works from public domain print editions means that no one owns a United States copyright in these works, so the Foundation (and you!) can copy and distribute it in the United States without permission and without paying copyright royalties. Special rules, set forth in the General Terms of Use part of this license, apply to copying and distributing Project Gutenberg-tm electronic works to protect the PROJECT GUTENBERG-tm concept and trademark. Project Gutenberg is a registered trademark, and may not be used if you charge for the eBooks, unless you receive specific permission. If you do not charge anything for copies of this eBook, complying with the rules is very easy. You may use this eBook for nearly any purpose such as creation of derivative works, reports, performances and research. They may be modified and printed and given away--you may do practically ANYTHING with public domain eBooks. Redistribution is subject to the trademark license, especially commercial redistribution.

*** START: FULL LICENSE ***

THE FULL PROJECT GUTENBERG LICENSE PLEASE READ THIS BEFORE YOU DISTRIBUTE OR USE THIS WORK

To protect the Project Gutenberg-tm mission of promoting the free distribution of electronic works, by using or distributing this work (or any other work associated in any way with the phrase "Project Gutenberg"), you agree to comply with all the terms of the Full Project Gutenberg-tm License (available with this file or online at http://gutenberg.net/license).

Section 1. General Terms of Use and Redistributing Project Gutenberg-tm electronic works

1.A. By reading or using any part of this Project Gutenberg-tm electronic work, you indicate that you have read, understand, agree to and accept all the terms of this license and intellectual property (trademark/copyright) agreement. If you do not agree to abide by all the terms of this agreement, you must cease using and return or

CHAPTER XXXV

destroy all copies of Project Gutenberg-tm electronic works in your possession. If you paid a fee for obtaining a copy of or access to a Project Gutenberg-tm electronic work and you do not agree to be bound by the terms of this agreement, you may obtain a refund from the person or entity to whom you paid the fee as set forth in paragraph 1.E.8.

1.B. "Project Gutenberg" is a registered trademark. It may only be used on or associated in any way with an electronic work by people who agree to be bound by the terms of this agreement. There are a few things that you can do with most Project Gutenberg-tm electronic works even without complying with the full terms of this agreement. See paragraph 1.C below. There are a lot of things you can do with Project Gutenberg-tm electronic works if you follow the terms of this agreement and help preserve free future access to Project Gutenberg-tm electronic works. See paragraph 1.E below.

1.C. The Project Gutenberg Literary Archive Foundation ("the Foundation" or PGLAF), owns a compilation copyright in the collection of Project Gutenberg-tm electronic works. Nearly all the individual works in the collection are in the public domain in the United States. If an individual work is in the public domain in the United States and you are located in the United States, we do not claim a right to prevent you from copying, distributing, performing, displaying or creating derivative works based on the work as long as all references to Project Gutenberg are removed. Of course, we hope that you will support the Project Gutenberg-tm mission of promoting free access to electronic works by freely sharing Project Gutenberg-tm works in compliance with the terms of this agreement for keeping the Project Gutenberg-tm name associated with the work. You can easily comply with the terms of this agreement by keeping this work in the same format with its attached full Project Gutenberg-tm License when you share it without charge with others.

1.D. The copyright laws of the place where you are located also govern what you can do with this work. Copyright laws in most countries are in a constant state of change. If you are outside the United States, check the laws of your country in addition to the terms of this agreement before downloading, copying, displaying, performing, distributing or creating derivative works based on this work or any other Project Gutenberg-tm work. The Foundation makes no representations concerning the copyright status of any work in any country outside the United States.

1.E. Unless you have removed all references to Project Gutenberg:

1.E.1. The following sentence, with active links to, or other immediate access to, the full Project Gutenberg-tm License must appear prominently whenever any copy of a Project Gutenberg-tm work (any work on which the phrase "Project Gutenberg" appears, or with which the phrase "Project Gutenberg" is associated) is accessed, displayed, performed, viewed, copied or distributed:

This eBook is for the use of anyone anywhere at no cost and with almost no restrictions whatsoever. You may copy it, give it away or re-use it under the terms of the Project Gutenberg License included with this eBook or online at www.gutenberg.net

1.E.2. If an individual Project Gutenberg-tm electronic work is derived from the public domain (does not contain a notice indicating that it is posted with permission of the copyright holder), the work can be copied and distributed to anyone in the United States without paying any fees or charges. If you are redistributing or providing access to a work with the phrase "Project Gutenberg" associated with or appearing on the work, you must comply either with the requirements of paragraphs 1.E.1 through 1.E.7 or obtain permission for the use of the work and the Project Gutenberg-tm trademark as set forth in paragraphs 1.E.8 or 1.E.9.

1.E.3. If an individual Project Gutenberg-tm electronic work is posted with the permission of the copyright holder, your use and distribution must comply with both paragraphs 1.E.1 through 1.E.7 and any additional terms imposed by the copyright holder. Additional terms will be linked to the Project Gutenberg-tm License for all works posted with the permission of the copyright holder found at the beginning of this work.

CHAPTER XXXV

1.E.4. Do not unlink or detach or remove the full Project Gutenberg-tm License terms from this work, or any files containing a part of this work or any other work associated with Project Gutenberg-tm.

1.E.5. Do not copy, display, perform, distribute or redistribute this electronic work, or any part of this electronic work, without prominently displaying the sentence set forth in paragraph 1.E.1 with active links or immediate access to the full terms of the Project Gutenberg-tm License.

1.E.6. You may convert to and distribute this work in any binary, compressed, marked up, nonproprietary or proprietary form, including any word processing or hypertext form. However, if you provide access to or distribute copies of a Project Gutenberg-tm work in a format other than "Plain Vanilla ASCII" or other format used in the official version posted on the official Project Gutenberg-tm web site (www.gutenberg.net), you must, at no additional cost, fee or expense to the user, provide a copy, a means of exporting a copy, or a means of obtaining a copy upon request, of the work in its original "Plain Vanilla ASCII" or other form. Any alternate format must include the full Project Gutenberg-tm License as specified in paragraph 1.E.1.

1.E.7. Do not charge a fee for access to, viewing, displaying, performing, copying or distributing any Project Gutenberg-tm works unless you comply with paragraph 1.E.8 or 1.E.9.

1.E.8. You may charge a reasonable fee for copies of or providing access to or distributing Project Gutenberg-tm electronic works provided that

- You pay a royalty fee of 20% of the gross profits you derive from the use of Project Gutenberg-tm works calculated using the method you already use to calculate your applicable taxes. The fee is owed to the owner of the Project Gutenberg-tm trademark, but he has agreed to donate royalties under this paragraph to the Project Gutenberg Literary Archive Foundation. Royalty payments must be paid within 60 days following each date on which you prepare (or are legally required to prepare) your periodic tax returns. Royalty payments should be clearly marked as such and sent to the Project Gutenberg Literary Archive

CHAPTER XXXV

Foundation at the address specified in Section 4, "Information about donations to the Project Gutenberg Literary Archive Foundation."

- You provide a full refund of any money paid by a user who notifies you in writing (or by e-mail) within 30 days of receipt that s/he does not agree to the terms of the full Project Gutenberg-tm License. You must require such a user to return or destroy all copies of the works possessed in a physical medium and discontinue all use of and all access to other copies of Project Gutenberg-tm works.

- You provide, in accordance with paragraph 1.F.3, a full refund of any money paid for a work or a replacement copy, if a defect in the electronic work is discovered and reported to you within 90 days of receipt of the work.

- You comply with all other terms of this agreement for free distribution of Project Gutenberg-tm works.

1.E.9. If you wish to charge a fee or distribute a Project Gutenberg-tm electronic work or group of works on different terms than are set forth in this agreement, you must obtain permission in writing from both the Project Gutenberg Literary Archive Foundation and Michael Hart, the owner of the Project Gutenberg-tm trademark. Contact the Foundation as set forth in Section 3 below.

1.F.

1.F.1. Project Gutenberg volunteers and employees expend considerable effort to identify, do copyright research on, transcribe and proofread public domain works in creating the Project Gutenberg-tm collection. Despite these efforts, Project Gutenberg-tm electronic works, and the medium on which they may be stored, may contain "Defects," such as, but not limited to, incomplete, inaccurate or corrupt data, transcription errors, a copyright or other intellectual property infringement, a defective or damaged disk or other medium, a computer virus, or computer codes that damage or cannot be read by your equipment.

1.F.2. LIMITED WARRANTY, DISCLAIMER OF DAMAGES - Except for the "Right of Replacement or Refund" described in paragraph 1.F.3, the Project Gutenberg Literary Archive Foundation, the owner of the Project Gutenberg-tm trademark, and any other party distributing a Project Gutenberg-tm electronic work under this agreement, disclaim all liability to you for damages, costs and expenses, including legal fees. YOU AGREE THAT YOU HAVE NO REMEDIES FOR NEGLIGENCE, STRICT LIABILITY, BREACH OF WARRANTY OR BREACH OF CONTRACT EXCEPT THOSE PROVIDED IN PARAGRAPH F3. YOU AGREE THAT THE FOUNDATION, THE TRADEMARK OWNER, AND ANY DISTRIBUTOR UNDER THIS AGREEMENT WILL NOT BE LIABLE TO YOU FOR ACTUAL, DIRECT, INDIRECT, CONSEQUENTIAL, PUNITIVE OR INCIDENTAL DAMAGES EVEN IF YOU GIVE NOTICE OF THE POSSIBILITY OF SUCH DAMAGE.

1.F.3. LIMITED RIGHT OF REPLACEMENT OR REFUND - If you discover a defect in this electronic work within 90 days of receiving it, you can receive a refund of the money (if any) you paid for it by sending a written explanation to the person you received the work from. If you received the work on a physical medium, you must return the medium with your written explanation. The person or entity that provided you with the defective work may elect to provide a replacement copy in lieu of a refund. If you received the work electronically, the person or entity providing it to you may choose to give you a second opportunity to receive the work electronically in lieu of a refund. If the second copy is also defective, you may demand a refund in writing without further opportunities to fix the problem.

1.F.4. Except for the limited right of replacement or refund set forth in paragraph 1.F.3, this work is provided to you 'AS-IS' WITH NO OTHER WARRANTIES OF ANY KIND, EXPRESS OR IMPLIED, INCLUDING BUT NOT LIMITED TO WARRANTIES OF MERCHANTIBILITY OR FITNESS FOR ANY PURPOSE.

1.F.5. Some states do not allow disclaimers of certain implied warranties or the exclusion or limitation of certain types of damages. If any disclaimer or limitation set forth in this agreement violates the law of the state applicable to this agreement, the agreement shall be

CHAPTER XXXV 290

interpreted to make the maximum disclaimer or limitation permitted by the applicable state law. The invalidity or unenforceability of any provision of this agreement shall not void the remaining provisions.

1.F.6. **INDEMNITY**

- You agree to indemnify and hold the Foundation, the trademark owner, any agent or employee of the Foundation, anyone providing copies of Project Gutenberg-tm electronic works in accordance with this agreement, and any volunteers associated with the production, promotion and distribution of Project Gutenberg-tm electronic works, harmless from all liability, costs and expenses, including legal fees, that arise directly or indirectly from any of the following which you do or cause to occur: (a) distribution of this or any Project Gutenberg-tm work, (b) alteration, modification, or additions or deletions to any Project Gutenberg-tm work, and (c) any Defect you cause.

Section 2. Information about the Mission of Project Gutenberg-tm

Project Gutenberg-tm is synonymous with the free distribution of electronic works in formats readable by the widest variety of computers including obsolete, old, middle-aged and new computers. It exists because of the efforts of hundreds of volunteers and donations from people in all walks of life.

Volunteers and financial support to provide volunteers with the assistance they need, is critical to reaching Project Gutenberg-tm's goals and ensuring that the Project Gutenberg-tm collection will remain freely available for generations to come. In 2001, the Project Gutenberg Literary Archive Foundation was created to provide a secure and permanent future for Project Gutenberg-tm and future generations. To learn more about the Project Gutenberg Literary Archive Foundation and how your efforts and donations can help, see Sections 3 and 4 and the Foundation web page at http://www.pglaf.org.

Section 3. Information about the Project Gutenberg Literary Archive Foundation

CHAPTER XXXV

The Project Gutenberg Literary Archive Foundation is a non profit 501(c)(3) educational corporation organized under the laws of the state of Mississippi and granted tax exempt status by the Internal Revenue Service. The Foundation's EIN or federal tax identification number is 64-6221541. Its 501(c)(3) letter is posted at http://pglaf.org/fundraising. Contributions to the Project Gutenberg Literary Archive Foundation are tax deductible to the full extent permitted by U.S. federal laws and your state's laws.

The Foundation's principal office is located at 4557 Melan Dr. S. Fairbanks, AK, 99712., but its volunteers and employees are scattered throughout numerous locations. Its business office is located at 809 North 1500 West, Salt Lake City, UT 84116, (801) 596-1887, email business@pglaf.org. Email contact links and up to date contact information can be found at the Foundation's web site and official page at http://pglaf.org

For additional contact information: Dr. Gregory B. Newby Chief Executive and Director gbnewby@pglaf.org

Section 4. Information about Donations to the Project Gutenberg Literary Archive Foundation

Project Gutenberg-tm depends upon and cannot survive without wide spread public support and donations to carry out its mission of increasing the number of public domain and licensed works that can be freely distributed in machine readable form accessible by the widest array of equipment including outdated equipment. Many small donations ($1 to $5,000) are particularly important to maintaining tax exempt status with the IRS.

The Foundation is committed to complying with the laws regulating charities and charitable donations in all 50 states of the United States. Compliance requirements are not uniform and it takes a considerable effort, much paperwork and many fees to meet and keep up with these requirements. We do not solicit donations in locations where we have not received written confirmation of compliance. To SEND DONATIONS or determine the status of compliance for any particular state visit http://pglaf.org

While we cannot and do not solicit contributions from states where we have not met the solicitation requirements, we know of no prohibition against accepting unsolicited donations from donors in such states who approach us with offers to donate.

International donations are gratefully accepted, but we cannot make any statements concerning tax treatment of donations received from outside the United States. U.S. laws alone swamp our small staff.

Please check the Project Gutenberg Web pages for current donation methods and addresses. Donations are accepted in a number of other ways including including checks, online payments and credit card donations. To donate, please visit: http://pglaf.org/donate

Section 5. General Information About Project Gutenberg-tm electronic works.

Professor Michael S. Hart is the originator of the Project Gutenberg-tm concept of a library of electronic works that could be freely shared with anyone. For thirty years, he produced and distributed Project Gutenberg-tm eBooks with only a loose network of volunteer support.

Project Gutenberg-tm eBooks are often created from several printed editions, all of which are confirmed as Public Domain in the U.S. unless a copyright notice is included. Thus, we do not necessarily keep eBooks in compliance with any particular paper edition.

Most people start at our Web site which has the main PG search facility:

http://www.gutenberg.net

This Web site includes information about Project Gutenberg-tm, including how to make donations to the Project Gutenberg Literary Archive Foundation, how to help produce our new eBooks, and how to subscribe to our email newsletter to hear about new eBooks.

General Science

CHAPTER XXXV

A free ebook from http://manybooks.net/

www.ingramcontent.com/pod-product-compliance
Lightning Source LLC
Chambersburg PA
CBHW050049230526
45470CB00004B/1461